# Focus in High School Mathematics: Reasoning and Sense Making in Algebra

*by*

Karen Graham
University of New Hampshire
Durham, New Hampshire

Al Cuoco
Education Development Center
Newton, Massachusetts

Gwen Zimmermann
Adlai E. Stevenson High School
Lincolnshire, Illinois

NATIONAL COUNCIL OF
TEACHERS OF MATHEMATICS

Library of Congress Cataloging-in-Publication Data

Graham, Karen (Karen J.)
  Focus in high school mathematics : reasoning and sense making in
algebra / by Karen Graham, Al Cuoco, Gwen Zimmermann.
     p. cm.
  Includes bibliographical references and index.
  ISBN 978-0-87353-640-0 (alk. paper)
  1.  Algebra--Study and teaching (Secondary)--United States. 2.
Mathematics--Study and teaching (Secondary)--United States.  I. Cuoco,
Albert. II. Zimmermann, Gwen. III. Title.
  QA159.G73 2010
  512.0071'2--dc22

                                   2009045939

The National Council of Teachers of Mathematics is a public voice of mathematics
  education, supporting teachers to ensure equitable mathematics learning of
  the highest quality for all students through vision, leadership, professional
                     development, and research.

Printed in the United States of America

# Table of Contents

# Preface

*Focus in High School Mathematics: Reasoning and Sense Making* (NCTM 2009) captures the direction for high school mathematics for students in the twenty-first century:

> Reasoning and sense making should occur in every mathematics classroom every day. In such an environment, teachers and students ask and answer such questions as "What's going on here?" and "Why do you think that?" Addressing reasoning and sense making does not need to be an extra burden for teachers struggling with students who are having a difficult time just learning the procedures. On the contrary, the structure that reasoning brings forms a vital support for understanding and continued learning. Currently, many students have difficulty because they find mathematics meaningless…. With purposeful attention and planning, teachers can hold all students in every high school mathematics classroom accountable for personally engaging in reasoning and sense making, and thus lead students to experience reasoning for themselves rather than merely observe it. (NCTM 2009, pp. 5–6)

This new publication urges a refocusing of the high school mathematics curriculum on reasoning and sense making, building on the guidelines for teaching and learning mathematics advocated by NCTM in *Principles and Standards for School Mathematics* (NCTM 2000). *Focus in High School Mathematics: Reasoning and Sense Making* makes the case that reasoning and sense making must reside at the core of all mathematics learning and instruction, at all grades. Moving forward from *Curriculum Focal Points for Prekindergarten through Grade 8 Mathematics* (NCTM 2006), *Focus in High School Mathematics: Reasoning and Sense Making* also addresses the need for the continuation of a coherent and well-articulated mathematics curriculum at the high school level.

The underlying principles of *Focus in High School Mathematics: Reasoning and Sense Making* are "reasoning habits" that should develop across the curriculum, along with "key elements" organized around five content strands. The book provides a group of examples that illustrate how these principles might play out in the classroom. Historically, NCTM has provided supplementary materials to accompany major publications that present official positions of the Council (e.g., the Teaching with Curriculum Focal Points series for *Curriculum Focal Points for Prekindergarten through Grade 8 Mathematics,* the Navigations Series for *Principles and Standards for School Mathematics*, the Addenda Series for *Curriculum and Evaluation Standards for School Mathematics* [NCTM 1989]). In keeping with this tradition, a series of supplementary books, Focus in High School Mathematics, provides additional guidance for ensuring that reasoning and sense making are part of the mathematics experiences of all high school students every day.

This series is intended for secondary mathematics teachers, curriculum specialists, mathematics supervisors, district administrators, and mathematics teacher educators. *Focus in High School Mathematics: Reasoning and Sense Making* underscores the critical role of the Process Standards outlined in *Principles and Standards* and provides a foundation for achieving the principal goals for the mathematical experiences of all secondary school students. Each volume in the Focus in High School Mathematics series presents detailed examples of worthwhile mathematical tasks, along with follow-up discussion. The examples and discussions are intended to help classroom teachers understand what it means to promote sense making and to find ways to increase it in their classrooms. The material could also be used as classroom cases in professional development. In addition, supervisors, curriculum specialists, and administrators might use the examples and discussions to catalyze conversations about shifts in the high school mathematics curriculum to bring them into alignment with the goals of *Focus in High School Mathematics: Reasoning and Sense Making*.

Although the books in the series focus on a particular content strand from *Principles and Standards* (e.g., geometry and measurement, algebra, statistics and probability), they are not intended to outline a curriculum for a particular content area. In fact, many of the examples in the books point to potential connections across content areas.

The authors of the present volume, *Focus in High School Mathematics: Reasoning and Sense Making in Algebra*, would like to acknowledge the following individuals for their critical review of drafts and technical help on the project: Anna Baccaglini-Frank, Bowen Kerins, Charlene Newton, and Kevin Waterman. Valuable advice also came from Dick Askey, Gail Burrill, John Carter, Roger Howe, Bill McCallum, and Glenn Stevens. Susan Forster provided useful assistance as a liaison from the writing group for *Focus in High School Mathematics: Reasoning and Sense Making.* In addition, the authors are grateful to all the teachers who offered opinions and reactions to the manuscript during the 2009 PROMYS for Teachers seminar session at Education Development Center, Newton, Massachusetts. The collective feedback from all of these individuals has made the final document stronger.

# General Introduction to the Focus in High School Mathematics Series

*Focus in High School Mathematics: Reasoning and Sense Making* addresses the need for reasoning to play a larger role in high school mathematics:

> A focus on reasoning and sense making, when developed in the context of strong content, will ensure that students can accurately carry out mathematical procedures, understand why those procedures work, and know how they might be used and their results interpreted…. Such a focus on reasoning and sense making will produce citizens who make informed and reasoned decisions, including quantitatively sophisticated choices about their personal finances, about which public policies deserve their support, and about which insurance or health plans to select. It will also produce workers who can satisfy the increased mathematical needs in professional areas ranging from health care to small business to digital technology. (NCTM 2009, p. 3)

*Focus in High School Mathematics: Reasoning and Sense Making* provides an outline for how reasoning and sense making might play out in core topic areas of the high school curriculum: numbers and measurement, algebra, geometry, and statistics and probability. The topics and examples contained in this publication and the supporting volumes do not represent an exhaustive list of topics that should be covered in any particular course or curriculum. The examples are meant to illustrate reasoning habits that all students at a variety of grade levels should know by the time they complete high school. As such, they provide multiple entry points for the students and, where appropriate, emphasize connections between several areas of mathematics. The discussions point to key teaching strategies that foster the development of reasoning and sense making. The strategies should be viewed as general and not tied to the particular context or task.

Most teachers and teacher educators would probably nod in agreement that reasoning and sense making are important to consider in the mathematical experiences of their students. However, the purpose of *Focus in High School Mathematics: Reasoning and Sense Making* and the Focus in High School Mathematics series is to highlight these as major goals of the study of secondary mathematics. Although reasoning and sense making may have been a part of secondary mathematics teaching and learning in the past, they are certainly worthy of being discussed in greater depth, and becoming a primary focus of our secondary mathematics teaching, in classrooms today. Therefore, with this shift in emphasis, it is important for NCTM to provide thoughtful examples of worthwhile tasks that can be pursued at a number of levels.

## The Role of Teaching

Often, high school mathematics teaching in the United States and Canada has been characterized by two main classroom activities; teachers share information, such as definitions of new terms and procedures for solving mathematics problems, and then students practice and perhaps discuss results of those procedures. Although these activities are important, such practices can lead to learning that is devoid of reasoning and sense making. By contrast, NCTM strongly supports a view of mathematics teaching and learning that focuses on reasoning, as described in *Mathematics Teaching Today* (NCTM 2007): "Teachers … must shift their perspectives about teaching from that of a process of delivering information to that of a process of facilitating students' sense making about mathematics" (p. 5).

A shift of perspective to one that views reasoning and sense making as primary goals for students' learning of mathematics will lead to a shift in choices made by the classroom teacher. For example, the teacher will choose tasks that allow students to see the need for sense making and provide opportunities for them to demonstrate their reasoning processes. Such tasks should also help students build on their informal knowledge of mathematics and see the logical connections with other areas of mathematics that they have learned. This shift may require changes in the structure of the classroom setting so that students are challenged and encouraged to explore mathematical situations both collaboratively and independently. Students should be expected to make conjectures and develop arguments to support them, connecting earlier knowledge with newly acquired knowledge.

As students are investigating and shaping ideas, they should have opportunities to interact directly and openly with one another and with the teacher. More details about the teacher's and students' roles in the classroom can be found in chapter 1, "Standards for Teaching and Learning," of *Mathematics Teaching Today*, which includes Standards describing characteristics of *worthwhile mathematical tasks* (Standard 3), components of a productive classroom *learning environment* (Standard 4), and suggestions for orchestrating mathematical *discourse* (Standard 5). The Focus in High School Mathematics series provides tasks, examples, and classroom vignettes that illustrate how a teacher might choose tasks and orchestrate classroom discourse to capitalize on student reasoning and promote sense making.

## The Role of Technology

Technology is integrated into the examples in these books in a strategic manner to enrich opportunities for students' reasoning and sense making. The power of recent technological tools (e.g., computer algebra systems, dynamic geometry software, and dynamic data representation tools) to enhance reasoning and sense making in mathematics is so great that it would be remiss to omit them from these volumes.

Increasingly, technology is an integral part of society and the research that is conducted in the majority of mathematics-related fields. We support the philosophy of *Focus in High School Mathematics: Reasoning and Sense Making* that "students can be challenged to take responsibility for deciding which tool might be useful in a given situation when they are allowed to choose from a menu of mathematical tools that includes technology. Students who have regular opportunities to discuss and reflect on how a technological tool is used effectively will be less apt to use technology as a crutch" (p. 14). The Focus in High School Mathematics series provides examples that show students using technology to reduce computational overhead, but the books also illustrate the use of technology in experimenting with mathematical objects and modeling mathematical structures.

## The Format of the Focus in High School Mathematics Series

*Focus in High School Mathematics: Reasoning and Sense Making* underscores the need to refocus the high school mathematics curriculum on reasoning and sense making. Companion books provide further insights into how these ways of thinking might develop in three major areas of content in high school mathematics:

- *Focus in High School Mathematics: Reasoning and Sense Making in Algebra*
- *Focus in High School Mathematics: Reasoning and Sense Making in Geometry*
- *Focus in High School Mathematics: Reasoning and Sense Making in Statistics and Probability*

The strand on reasoning and sense making with numbers and measurement discussed in *Focus in High School Mathematics: Reasoning and Sense Making* receives primary attention in *Focus in High School Mathematics: Reasoning and Sense Making in Geometry,* but aspects of this strand are also addressed in the other two content books.

## Reasoning Habits

To detail what mathematical reasoning and sense making should look like across the high school curriculum, *Focus in High School Mathematics: Reasoning and Sense Making* provides a list of "reasoning habits." The intent is not to present a new list of topics to be added to the high school curriculum: "Approaching the list as a new set of topics to be taught in an already crowded curriculum is not likely to have the desired effect. Instead, attention to reasoning habits needs to be integrated within the curriculum to ensure that students both understand and can use what they are taught" (p. 9). The reasoning habits are described and illustrated in the examples throughout the companion books in the Focus in High School Mathematics series.

## Key Elements

*Focus in High School Mathematics: Reasoning and Sense Making* identifies "key elements" for each of the strands. These key elements are intended to provide "a lens through which to view the potential of high school programs for promoting mathematical reasoning and sense making" (p. 18).

## Content Expectations

As *Focus in High School Mathematics: Reasoning and Sense Making* suggests, readers wishing for more detailed content recommendations should refer to chapter 7, "Standards for Grades 9–12," of *Principles and Standards for School Mathematics* (NCTM 2000). However, for the readers' convenience, each companion volume shows the grades 9–12 expectations of the relevant Standard (Algebra, Geometry, or Data Analysis and Probability) in the appendix, along with the grades 9–12 expectations for the Number and Operations and the Measurement Standards, which are addressed by all three volumes.

# Introduction
## to *Focus in High School Mathematics: Reasoning and Sense Making in Algebra*

Algebra and algebraic reasoning remain at the core of the high school mathematics curriculum. Algebra's dominance in the school curriculum is related to the importance of algebra in more advanced areas of mathematics, the usefulness of algebraic reasoning in all walks of life, and the role of algebra as a tool for the mathematical modeling required in many technological and scientific fields. Algebra's vital role in the school curriculum is reflected in many recent policy documents (e.g., NCTM 2000; Mathematical Association of America 2007; National Mathematics Advisory Panel 2008), as well as in the often-heated debate over whether algebra should be a required course for all eighth-grade students. The executive order that established the National Mathematics Advisory Panel places algebra at the top of the list of topics for the panel to consider, identifying as item (*a*) "the critical skills and skill progression for students to acquire competence in algebra and readiness for higher levels of mathematics" (National Mathematics Advisory Panel 2008, p. 7). The Common Core State Standards for K–12 Mathematics, developed under the leadership of the National Governors Association Center for Best Practices and the Council of Chief State School Officers, provide a strong foundation for algebra, with a focus in early grades on number and operations and in middle grades on ratio and proportional reasoning, supporting high school algebra standards that include reasoning and sense making in addition to procedural fluency. *Principles and Standards for School Mathematics*, published by the National Council of Teachers of Mathematics in 2000, and the recent report *Algebra: Gateway to a Technological Future,* released by the Mathematical Association of America in 2007, outline key standards for algebra and the connections between algebra and other areas of mathematics.

Although formal algebra coursework begins in grade 9 for most students, many recent reports (NCTM 2000; National Mathematics Advisory Panel 2008; Greenes 2008) stress the importance of laying the foundation for algebra and algebraic thinking as early as kindergarten. The building blocks for algebra include fluency with whole numbers and fractions, experience in analyzing properties of patterns and shapes and using proportional reasoning, and several of the algebraic habits that we describe in this book. Many of the early experiences involve the use of manipulatives and concrete models. It is then the job of the teachers of beginning algebra to "proceed carefully, ensuring that their students have some images that will give meaning and coherence while weaning them away from concrete models and examples so that they can exploit the flexibility and breadth of vision that formal algebraic procedures bring" (Barbeau and Brown 1997). This book provides examples of how teachers can build on students' concrete experiences and help them to make connections between these experiences and the more formal aspects of algebra, which are the focus of higher-level mathematics.

## Organizing Frameworks for School Algebra

Usiskin (1988, pp. 11–16) describes several conceptions of algebra that he says have influenced the development of the school algebra curriculum and give meaning to purposes for teaching school algebra:

**Conception 1:** Algebra as generalized arithmetic—variables are pattern generalizers, and key instructions to students are *translate* and *generalize*.

**Conception 2:** Algebra as a study of procedures for solving certain kinds of problems—variables are unknowns or constants, and key instructions are *simplify* and *solve*.

**Conception 3:** Algebra as the study of relationships among quantities—variables are arguments or parameters, and key instructions are contained in questions such as "What happens to the value of $1/x$ as $x$ gets larger and larger?"

**Conception 4:** Algebra as the study of structures—variables are marks on paper, and key instructions are contained in commands such as "Factor the polynomial $3x^2 + 4ax - 132a^2$."

Other mathematicians and mathematics educators have outlined similar conceptions of school algebra (Bass 1998; Mathematical Sciences Research Institute 2008; Dossey 1998; Fey and Good 1985; Thompson 2008). In particular, Dossey (1998) describes the different conceptions or models for school as structural, focusing on functions and relations, building models, and linguistic (p.18). He argues that aspects of each conception are necessary in practice and that none of the approaches can survive on its own: "What we really have to do is to think of how to merge them to support a coherent program of the study of algebra with four main goals" (p. 19). The Standard for algebra outlined in *Principles and Standards* supports this view. The elements of the Algebra Standard that cut across pre-K–grade 12 echo these conceptions:

Instructional programs from prekindergarten through grade 12 should enable all students to—

- Understand patterns, relations, and functions

- Represent and analyze mathematical situations and structures using algebraic symbols

- Use mathematical models to represent and understand quantitative relationships

- Analyze change in various contexts. (NCTM 2000, p. 37).

We believe that an emphasis on reasoning and sense making is essential to students' success in algebra, no matter what conception or approach instruction takes. As *Navigating through Reasoning and Proof in Grades 9–12*, a volume in the recent Navigations Series, states, "instruction that emphasizes reasoning can transform algebra from a procedural and formula-based study, focused on the ability to memorize, apply, and combine processes, to a study that calls for creative and original thinking" (Burke et al. 2008, p. 18). Algebra continues to be an important tool—one that is required in many disciplines and careers—and an emphasis on reasoning and sense making will help students appreciate its value and understand why it continues to be an integral part of the high school curriculum.

# Key Elements of Algebraic Reasoning

*Focus in High School Mathematics: Reasoning and Sense Making* (NCTM 2009) discusses the key elements for reasoning and sense making with algebra in two parts: reasoning and sense making with algebraic symbols, and reasoning and sense making with functions. Key elements of reasoning and sense making with algebraic symbols include:

- *Meaningful use of symbols.* Choosing variables and constructing expressions and equations in context; interpreting the form of expressions and equations; manipulating expressions so that interesting interpretations can be made.

- *Mindful manipulation.* Connecting manipulation with the laws of arithmetic; anticipating the results of manipulations; choosing procedures purposefully in context; picturing calculations mentally.

- *Reasoned solving.* Seeing solution steps as logical deductions about equality; interpreting solutions in context.

- *Connecting algebra with geometry.* Representing geometric situations algebraically and algebraic situations geometrically; using connections in solving problems.

- *Linking expressions and functions.* Using multiple algebraic representations to understand functions; working with function notation. (NCTM 2009, p. 31)

Key elements of reasoning and sense making with functions include the following:

- *Using multiple representations of functions.* Representing functions in various ways, including tabular, graphic, symbolic (explicit and recursive), visual, and verbal; making decisions about which representations are most helpful in problem-solving circumstances; and moving flexibly among those representations.

- *Modeling by using families of functions.* Working to develop a reasonable mathematical model for a particular contextual situation by applying knowledge of the characteristic behaviors of different families of functions.

- *Analyzing the effects of parameters.* Using a general representation of a function in a given family (e.g., the vertex form of a quadratic, $f(x) = a(x - h)^2 + k$), to analyze the effects of varying coefficients or other parameters; converting between different forms of functions (e.g., the standard form of a quadratic and its factored form) according to the requirements of the problem-solving situation (e.g., finding the vertex of a quadratic or finding its zeros). (NCTM 2009, p. 41)

Each chapter in this book deals with one or more of these key elements of algebraic reasoning in greater detail. In addition to the key elements, certain habits of mind are particularly evident in algebraic reasoning.

## Specific Habits of Mind in Algebraic Reasoning

In the examples developed in this book, we have noted several algebraic sub-themes within the reasoning habits presented in *Focus in High School Mathematics: Reasoning and Sense Making.* Table 0.1 shows the habits of mind that we believe are of particular importance in algebraic reasoning.

Table 0.1
*Habits of Mind in Reasoning and Sense Making in Algebra*

---

**Analyzing a problem,** *for example,*

- *defining relevant variables and conditions* carefully, including units if appropriate;
- *seeking patterns and relationships;*
- *looking for hidden structure* (for example, finding equivalent forms of expressions that reveal different aspects of a problem).

---

**Implementing a strategy,** *for example,*

- *making purposeful use of procedures;*
- *monitoring progress toward a solution,* including reviewing a chosen strategy and other possible strategies generated by oneself or others.

---

*(Continued on the next page)*

Table 0.1

*Habits of Mind in Reasoning and Sense Making in Algebra—Continued*

**Reflecting on a solution to a problem,** *for example,*

- *interpreting a solution* and how it answers the problem, including making decisions under uncertain conditions;
- *considering the reasonableness of a solution*, including whether any numbers are reported to an unreasonable degree of accuracy;
- *generalizing a solution* to a broader class of problems and looking for connections to other problems.

## Overview of This Book

Each of the chapters in this book provides examples of how reasoning and sense making might play out in the high school algebra curriculum. Chapter 1 looks at the ways in which algebra and geometry "talk to each other" about area by exploring three specific contexts: finding area formulas for polygons, maximizing area and connecting it to the arithmetic-geometric mean inequality, and Heron's formula for the area of triangles. Chapter 2 explores the habit of seeking and expressing regularity in repeated calculations in three contexts: word problems, fitting lines to data, and monthly loan payments. Chapter 3 presents ideas about how students might make sense of the algebra of formal expressions in the high school curriculum by exploring the process of factoring and completing the square, combinatorial phenomena, and patterns in the factors of a sequence of polynomials. The epilogue surveys the journey that the readers have made, inviting them to reflect on the examples and classroom vignettes contained in the book and to consider how to make reasoning and sense making integral parts of the high school curriculum and classroom practice.

The examples in each chapter illustrate a range of content and grade levels across the 9–12 spectrum and reflect essential concepts in algebra that the authors believe students should have experience with before graduating from high school. The concepts that are the focus of the examples could arise in algebra courses or other courses, such as geometry or probability and statistics.

Again, readers should note that this book is not an algebra curriculum but instead contains key algebraic ideas that illustrate reasoning and sense making that should be part of any mathematics curriculum in grades 9–12. The examples do not in any way represent an exhaustive list of topics. Readers can refer to *Principles and Standards for School Mathematics* (NCTM 2000) and the references at the end of the book for more detail on recommended goals for school algebra and other related resources.

# Chapter 1

# Algebra and Geometry

Students can develop ideas about algebra and geometry together throughout the high school curriculum, enriching their study of each. Geometric interpretations of algebraic identities can help them give meaning to and make sense of algebraic symbols and calculations. Conversely, casting geometric phenomena in algebraic terms can give them a way to reason about the geometry, leading to interesting and nontrivial geometric conjectures and their proofs.

Several topics in the high school curriculum invite students to make connections very naturally between algebra and geometry—analytic geometry, vector methods, geometric probability, and conic sections are examples. This chapter looks at the ways in which algebra and geometry "talk to each other" about *area* in the following specific contexts:

1. Finding area formulas for polygons

2. Maximizing area and connecting this work to the arithmetic-geometric mean inequality

3. Working with Heron's formula for the area of a triangle

Activities in these contexts can highlight many of the reasoning habits and key elements of algebra outlined in *Focus in High School Mathematics: Reasoning and Sense Making* (NCTM 2009).

## Area Formulas

Most students come to high school knowing that the formula for the area of a rectangle is "area equals base times height." Many teachers revisit the formula briefly to make sure that students understand why it makes sense. One way to do this is by generic example. Consider a rectangle that is 3 units by 5 units and can be partitioned into 15 squares, each with area 1 square unit (see fig. 1.1).

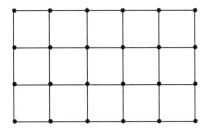

Fig. 1.1. A rectangle that is 3 units by 5 units, partitioned into unit squares

Analyzing the area of a rectangle by partitioning it does not work if no common unit of measure goes evenly into the lengths of the base and height. For example, the method doesn't work for a rectangle whose dimensions are 3 and $\sqrt{2}$, because 3 and $\sqrt{2}$ are not *commensurable*. Extending this method for obtaining the $A = bh$ formula to rectangles like this requires a continuity argument.

This method of partitioning works fine for all rectangles whose side lengths are integers in some measurement system (or, put another way, when there is a unit length that divides the lengths of the base and height evenly). Recognizing this fact is probably enough to convince most students to take the formula as an axiom for area.

Once the class accepts the formula, students can begin to explore the geometry-algebra connection. For example, they can interpret the algebraic fact that

$$2(bh) = (2b)h = b(2h)$$

geometrically by observing that the area of a rectangle doubles if they double the length of the base and keep the height the same *or* if they double the length of the height and keep the base the same, as figure 1.2 illustrates.

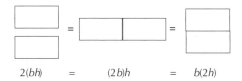

$$2(bh) \quad = \quad (2b)h \quad = \quad b(2h)$$

Fig. 1.2. Doubling the area of a rectangle by doubling its measurement in one direction

This simple example already exhibits some of the basic properties of area:

- Translating, rotating, or reflecting a figure does not change its area (area is invariant under rigid motions; for example, duplicating a rectangle doubles its area).

- Cutting up a figure into a finite number of pieces and rearranging them does not change the area (area is invariant under finite dissections).

These properties of area are essential in some advanced treatments of area that provide, for example, rigorous proofs of the fact that the area of a rectangle is $bh$, even if $b$ and $h$ are not commensurable.

Note that *figure* in these cases signifies a polygon or a circle. These area-preserving transformations allow one to keep track of changes in area in the process of moving figures around, duplicating them, or cutting them up and rearranging the pieces. A major goal of this chapter is to show how these actions are reflected in algebraic transformations of the area formulas and, conversely, how transformations of the area formulas are reflected in geometric transformations of the figures.

## The trapezoid

Imagine a class in which the students know the formulas for the areas of a rectangle and a parallelogram. How can they use this information to develop the formula for the area, $A$, of a trapezoid (see fig. 1.3)?

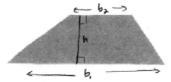

Fig. 1.3. A trapezoid with height $h$ and bases $b_1$ and $b_2$

Directing students to work independently on a problem like this allows them to exercise their creativity and then showcase their methods for the class. The experience also encourages them to make some meaningful interpretations of equivalent algebraic expressions.

One way to determine the area of a trapezoid by using the other, simpler formulas is to turn the trapezoid into a parallelogram of equal area, keeping track of changes in the linear dimensions in the process. Consider three methods that have come up in various classrooms:

> Developing the formula for the area of a trapezoid from the formulas for the area of a rectangle and a parallelogram provides students with an opportunity to exercise many of the reasoning habits described in *Focus in High School Mathematics: Reasoning and Sense Making*.

1. Duplicate the trapezoid and arrange two trapezoids to make a parallelogram with twice the area, as in figure 1.4.

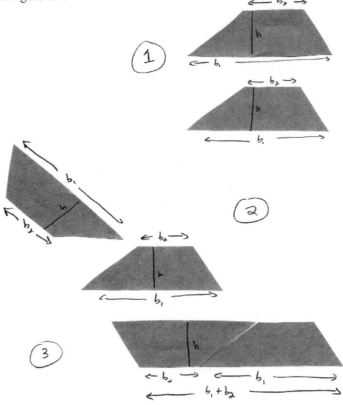

Fig. 1.4. Duplicating a trapezoid and arranging the two trapezoids to make a parallelogram

Students who use this argument might reason that, because the area of the parallelogram is $2A$ and the area of a parallelogram is equal to its base times its height,

$$2A = (b_1 + b_2)h,$$

so,

$$A = \frac{1}{2}\left[(b_1 + b_2)h\right].$$

> All the transformations made in the process of finding the area of a trapezoid call for justification. For instance, is the resulting figure in example 1 really a parallelogram? Why? Depending on your class, you could simply take note of a need for justification and save the justification itself for later, or you could include its development as part of the activity.

2. Slice the trapezoid along its midline and arrange the pieces to make a parallelogram of equal area, as in figure 1.5.

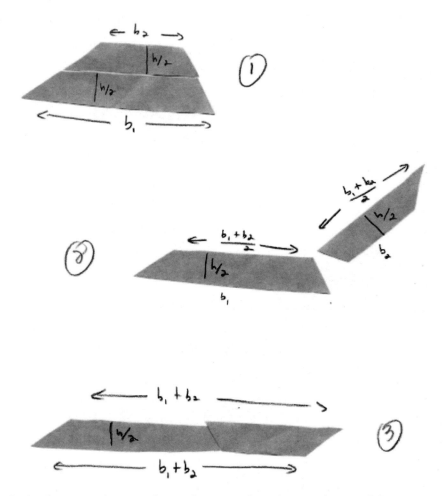

Fig. 1.5. Cutting the trapezoid on its midline and arranging the pieces to make a parallelogram of equal area

The height of the resulting parallelogram is half the height of the trapezoid, and the parallelogram's base is the sum of the lengths of the trapezoid's bases. The parallelogram has the same area as the area of the trapezoid. So, a student who uses this line of reasoning can express the area in this way:

$$A = (b_1 + b_2)\left(\frac{h}{2}\right)$$

> Quite a few geometric results are required to show that the method in example 2 works. For instance, the length of the midline of a trapezoid must be half the sum of the bases, and the midline must bisect the trapezoid's height.

3. Slice the trapezoid along its midline and then cut off two triangles, arranging the pieces to make a rectangle of equal area, as in figure 1.6.

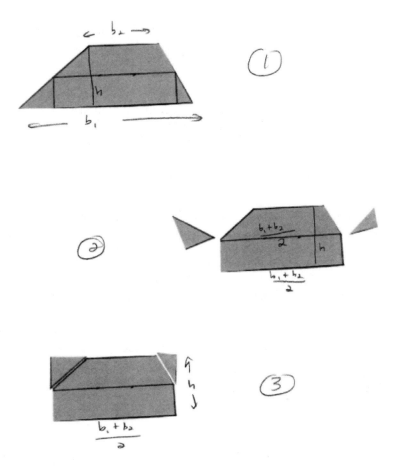

Fig. 1.6. Slicing a trapezoid along its midline, cutting off two triangles, and arranging the pieces to make a rectangle

Students who use the method in example 3 need to make sure that the resulting figure really is a rectangle.

The trapezoid and the rectangle have the same area and the same height, but the base of the rectangle is congruent to the midline of the trapezoid, so its length is half the sum of the bases of the trapezoid. Students who use this dissection might express the area of the trapezoid in the following way:

$$A = \left(\frac{b_1 + b_2}{2}\right) h$$

Thus, a class in which students are reasoning from the formulas for the areas of a rectangle and a parallelogram to develop a formula for the area, $A$, of a trapezoid can come up with at least three expressions for $A$:

$$\frac{1}{2}\left[(b_1 + b_2)h\right], \qquad (b_1 + b_2)\left(\frac{h}{2}\right), \qquad \text{and} \qquad \left(\frac{b_1 + b_2}{2}\right) h$$

What is the additional value of this exercise in which students generate these three expressions for the area of a trapezoid, compared with that of a lesson in which they develop or learn a single expression?

Algebraically, these three expressions are all equivalent—and many algebra students can get from one to another by using the basic rules for calculating with algebraic expressions. This result implies that the expressions produce the same value for any choice of numbers $b_1$, $b_2$, and $h$. Is anything more to be learned?

This experience provides students with an opportunity to explore the fact that algebraically equivalent expressions define the same *function*. Especially for beginning students, relating variables to numerical quantities (like length and area) can infuse algebraic expressions with meaning.

Interpreting different forms of the same area formula as different sets of instructions is an activity that teachers can carry out informally—in a beginning algebra class, for example—or with more justification required as students encounter some of the formal results in geometry. Below, we offer ideas to try, and we follow these suggestions with a look inside the classroom.

1.  In classrooms where we've seen activities like this in action, students don't usually start with the task of showing that the expressions are algebraically equivalent. Typically, different algorithms for calculating the same thing emerge from different lines of reasoning in the class—in this example, the same area can be calculated in several different ways. In these situations, algebra can be the perfect tool for showing that the different-looking expressions will always produce the same result when the same values are substituted for the variables in each.

2.  Once students are used to this idea of calculating area in more than one way, you can turn the tables and ask them to come up with geometric interpretations of expressions that are algebraically equivalent but that describe different calculations. For example, you might present different forms for the formula for the area of a triangle:

Consider an interesting image that you might use to bring closure to this triangle activity: Take each of the area calculations for the trapezoid and ask what happens to the geometry and to the algebra as $b_2$ approaches 0.

$$\frac{1}{2}bh, \qquad \left(\frac{b}{2}\right)h, \qquad b\left(\frac{h}{2}\right)$$

Ask students to pick one and show how a triangle can be moved, copied, or dissected to get a parallelogram whose area calculation mimics the chosen expression.

### In the classroom

The vignette below presents a hypothetical geometry class—a composite of classes that we've taught or visited over the years—so our hypothetical students possess some basic facts about angles and parallel lines. In a less formal setting, all that is required is a working knowledge of area formulas for parallelograms and an intuitive understanding of the idea that area is invariant under rigid motions. The class is looking for a formula for the area of a trapezoid, given the length of both bases and the height.

*Tom:* Remember, yesterday we found that the area of a parallelogram was base times height. We cut off a triangle from the end and moved it so we had a rectangle. Maybe we can do something like that.

*Jane:* If we cut off one of the end triangles and slide it over to the other end, it doesn't fit perfectly. That doesn't work. It would work if the trapezoid were an isosceles trapezoid. Then when we cut one end triangle, we could rotate and slide it to the other end, and the area of the figure is base times height because we get a rectangle.

*Dick:* Mrs. Euclid gave us two copies of the same trapezoid. Maybe we should try using both trapezoids. The only shapes we know how to find the area for are a rectangle and a parallelogram, so can we make the trapezoid into one of those shapes?

*Jane:* Wait! I think I've got it. If we match up the leg in the first trapezoid with the same leg in the second copy, I think we get something. We have to turn the trapezoid. Look, we get a parallelogram.

*Tom:* How do you know it's a parallelogram?

*Jane:* Well, the top and bottom sides are $b_1$ and $b_2$ added together, so that pair of opposite sides is congruent, and since the two other sides are actually the same legs from the original trapezoid, they must be the same. So we know that both pairs of sides are congruent—it has to be a parallelogram.

*Dick:* How does this help us figure out the area of the trapezoid?

> What would you do if your students had not yet learned the method that Jane describes for determining that a quadrilateral is a parallelogram?

*Tom:* I see it! We have a parallelogram, and we know the area for a parallelogram is base times height. In our parallelogram, the base comes from the trapezoid. So the base is $b_1 + b_2$. And the height of the parallelogram is the same as the height of the trapezoid, which is $h$. Then I multiply those to get the area, $(b_1 + b_2)h$.

*Dick:* But the trapezoid is only half the figure, so the area for the trapezoid is $\frac{1}{2}(b_1 + b_2)h$.

*Mrs. Euclid:* Very nice. But I wonder—how do you know that the top and bottom bases of the parallelogram are really segments? You made them from two segments—the top and bottom of the trapezoid. Maybe there's a bend where you put them together.

*[The students look at their work. Jane draws on the diagram (see fig. 1.7).]*

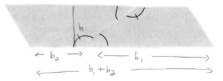

Fig 1.7. A parallelogram with Jane's markings

> In a class that comes before work with these geometric results about parallel lines, Mrs. Euclid would probably just flag the concern about whether the base segments are straight and see how the students react.

*Jane:*      Look, the angles at either end of the same leg of the trapezoid are supplementary because the bases are parallel, and we have a theorem about that. When the trapezoids are put together, the supplementary angles get put next to each other, so the place where the bases meet makes a 180° angle.

*Tom:*      So, we're good.

*Mrs. Euclid:*      No arguments here. Great work.

## Maximizing Area

It is common in many algebra classes to use quadratic functions to show that among all rectangles of a given perimeter, the square maximizes the area. Another way to derive that fact highlights the algebra-geometry interplay. Imagine a $12 \times 16$ rectangle (see fig. 1.8).

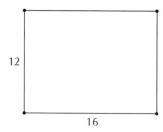

Fig. 1.8. A rectangle that is 12 units by 16 units

The basic idea is to dissect a rectangle so that it fits—with something left over—into a square with the same perimeter. Once again, setting a class to work on this problem is likely to produce several different approaches that will lead to some interesting algebraic identities.

One way to reason through the problem might be as follows. The rectangle's perimeter is 56, so a square with the same perimeter has side 14:

$$14 = \frac{16+12}{2}$$

We can cut a $2 \times 12$ rectangle from the big rectangle with a vertical slice at the rightmost end. Then we can slide the skinny rectangle up, rotate it 90° counterclockwise, and slide it over on top of what is left of the big rectangle, until the side of length 2 aligns with the side of length 12. The result is shown in figure 1.9. The original rectangle has thus been rearranged (the gray pieces) to fit inside a $14 \times 14$ square, with a $2 \times 2$ square left over.

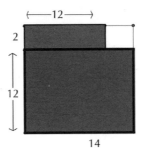

Fig. 1.9. A square that results from rearranging the pieces of the original rectangle

A useful algebraic habit is to *delay evaluation* of numerical expressions, to see where they came from. The following reasoning provides an example. The side length of the square is 14. This is obtained by averaging the side lengths of the rectangle:

$$\frac{16+12}{2}$$

The leftover "hole" is a $2 \times 2$ square, but the 2 came from $14 - 12$, which came from

$$\frac{16+12}{2} - 12 = \frac{16-12}{2}.$$

If we express the area of the rectangle as the difference of the areas of the big and small squares, we have a numerical identity:

$$\left(\frac{16+12}{2}\right)^2 - \left(\frac{16-12}{2}\right) = 16 \cdot 12$$

*Focus in High School Mathematics: Reasoning and Sense Making* refers to this habit of delaying evaluation of numerical expressions as "mindful manipulation."

Nothing is special about 16 and 12 here. If the dimensions of the rectangle are $a$ and $b$ (with $a \geq b$), the same reasoning shows that

$$\left(\frac{a+b}{2}\right)^2 - \left(\frac{a-b}{2}\right)^2 = ab. \qquad (*)$$

This identity shows that a rectangle of maximum area with fixed perimeter is a square, and it indicates just how far off the rectangle is from its maximum area.

Students can read a great deal from the identity (*). For example, it says that the product $ab$ is always less than the square of the average $\frac{a+b}{2}$ by a specific amount. When is the "error" term $\left(\frac{a-b}{2}\right)^2$ equal to 0? What does this say about the geometry?

Identity (*) has another geometric application. In many geometry courses, students investigate the *arithmetic-geometric mean inequality*. This inequality states that if $a$ and $b$ are nonnegative real numbers, their arithmetic mean is never smaller than their geometric mean; in symbols,

$$\frac{a+b}{2} \geq \sqrt{ab} ,$$

with equality if and only if $a = b$.

A simple proof of the arithmetic-geometric mean inequality uses similar triangles and the fact that an angle inscribed in a semicircle is a right angle (see fig. 1.10).

The arithmetic mean for $a$ and $b$ is the number $d$ so that $a$, $d$, and $b$ form an arithmetic sequence: $d - a = b - d$. The geometric mean is the number $r$ so that $a$, $r$, and $b$ form a geometric sequence:

$$\frac{a}{r} = \frac{r}{b}.$$

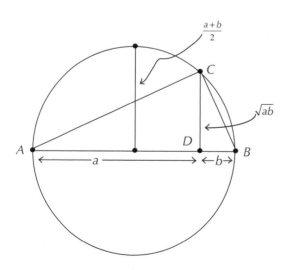

Fig. 1.10. An angle inscribed in a semicircle

Identity (*) provides another perspective on the arithmetic-geometric mean inequality. Because squares are never negative, the equality

In figure 1.10, $\triangle ACD \sim \triangle CBD$, so $\dfrac{AD}{CD} = \dfrac{CD}{BD}$. The arithmetic-geometric mean inequality follows from this relationship.

$$\left(\frac{a+b}{2}\right)^2 - \left(\frac{a-b}{2}\right)^2 = ab$$

implies that

$$\left(\frac{a+b}{2}\right)^2 \geq ab, \qquad (**)$$

with equality precisely when $a = b$. If $a$ and $b$ are nonnegative, both sides of (**) are nonnegative, and hence, we can take square roots and preserve the inequality:

$$\frac{a+b}{2} \geq \sqrt{ab}$$

This is the arithmetic-geometric mean inequality.

Again, consider the value of going through this reasoning process and what it adds to learning. The circle argument that establishes the arithmetic-geometric mean inequality is simple and elegant. What is gained by incorporating the additional perspective provided by identity (*)?

1.  The circle argument depends on results about inscribed angles and similar triangles. The algebraic approach afforded by (*) requires much less background. An alternative method of development to this approach involves turning the tables again and asking first-year algebra students to prove (*) by using the basic rules of algebra, without any reference to squares and rectangles, and *then* bringing in the geometry.

2.  As we developed the identity (*) here, it emerges as a way to express the general result that a square maximizes the area of all rectangles with a fixed perimeter. The fact that the inequality can also be used for another application is one of the wonderful features of algebraic identities: they often give you more than you ask for.

3.  The approach generalizes. For example, students in the upper grades could be asked to generalize the proof to higher dimensions. This work could lead to an investigation that might proceed along the lines indicated in the next section.

## Extending the idea

There is an arithmetic-geometric mean inequality in any number of variables. The $n$th power of the arithmetic mean of $n$ nonnegative real numbers is never smaller than the product of the numbers. For example, students accustomed to reasoning and sense making might wonder if there is an arithmetic-geometric inequality for three variables. They can use numerical experiments to shape a conjecture: If $a$, $b$, and $c$ are nonnegative numbers, then

$$\left(\frac{a+b+c}{3}\right)^3 \geq abc.$$

This is in fact true. How might students prove it?

Two questions arise from extending the investigation of the arithmetic-geometric mean inequality to higher dimensions:

1. Does this inequality have a geometric justification?

2. Can we find an expression for the "error" in the inequality?

> The *error* is what you have to subtract from the left-hand side of the inequality to get a quantity that equals the right-hand side.

Because the students have three variables, both of these questions suggest that they look at a box whose unequal dimensions are *a, b,* and *c,* with $a > b > c$. Extending the idea from two dimensions, they might plan to dissect the box into pieces that can fit completely inside a cube with a side length of $L = \frac{a+b+c}{3}$. Such a cube would have the same edge perimeter as the $a \times b \times c$ box. Why?

Such an investigation can lead to surprising and complex algebraic identities. For example, if $L = \frac{a+b+c}{3}$, then either $a > b > L > c$ or $a > L > b > c$. Students might be asked to prove this. One way for them to start is to assume that $a > b > L > c$. Then, by using the above notation, they have

> The *edge perimeter* is the sum of the lengths of the 12 edges of the box.

$$L^3 - (L-b)^2c - (L-b)(L-c)c - (L-c)^2L = abc.$$

As in the two-variable case, this result comes from decomposing an $a \times b \times c$ box into a cube of side $\frac{a+b+c}{3}$ with room to spare; the three expressions that are subtracted from the volume of a box show the excess of the volume of the cube over the volume of the box. Expanding the equation above after replacing $L$ by its value produces

$$\left(\frac{a+b+c}{3}\right)^3 - \left(\frac{a-2b+c}{3}\right)^2 c - \left(\frac{a-2b+c}{3}\right)\left(\frac{a+b-2c}{3}\right)c - \left(\frac{a+b-2c}{3}\right)^2\left(\frac{a+b+c}{3}\right) = abc.$$

We leave the actual geometric dissections that lead to identities like this as a challenge for you and your students. One way to get started is to use numerical values for *a, b,* and *c,* delaying the evaluation of expressions until the form of the calculations is visible. Styrofoam models can also be useful for building the actual dissection.

Two related algebra-geometry questions can be used for further investigation with students:

1. Can you find the dimensions of a rectangle if you know its perimeter and area? If so, how? If not, what else would you need to know?

2. Can you find the dimensions of a rectangular box if you know its edge perimeter and volume? If so, how? If not, what else would you need to know?

## Heron's Formula

One of the most beautiful theorems in geometry, named after Heron of Alexandria, gives a formula for the area of a triangle in terms of the lengths of its three sides.

*Heron's formula:* If the sides of a triangle have lengths $a$, $b$, and $c$, the area, $A$, of the triangle is

$$A = \sqrt{s(s-a)(s-b)(s-c)},$$

where $s = \dfrac{1}{2}(a+b+c)$.

Proving Heron's formula is a good exercise for advanced high school students. A proof is given at the end of this section; however, even without a proof, students can explore some reasons why the formula makes sense. As a first step, you might ask students to replace $s$ by $\dfrac{a+b+c}{2}$ and show that Heron's formula can also be written in a symmetric form:

$$A = \frac{1}{4}\sqrt{(a+b+c)(a+b-c)(a+c-b)(b+c-a)} \qquad (***)$$

This is a nice exercise in algebraic calculation, and a computer algebra system can be used either to generate the simplification or to verify it, as illustrated in figure 1.11.

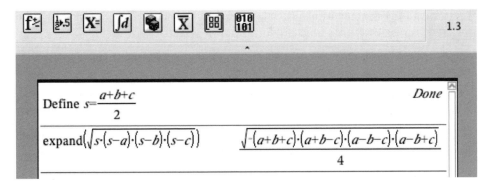

Fig. 1.11. A CAS screenshot that displays the algebraic calculation when a given substitution for $s$ is made in Heron's formula

Some additional problems that you might ask students to think about include the following:

> The CAS used in producing figure 1.11 gives a slightly different expression from the one in (***). Such discrepancies often occur with technology, and the experience can provide a good opportunity for students to think about equivalent expressions.

1. Use Heron's formula to find the area of an equilateral triangle whose side has length 10. Check your result against the result of the formula $A = \dfrac{1}{2}bh$.

2. What is the area of a "triangle" whose sides have length 4, 5, and 9? What happens when you apply Heron's formula to these numbers? Under what conditions on $a$, $b$, and $c$ will Heron's formula produce 0, and what do these "triangles" look like?

3. We could think of Heron's formula as a function of three inputs:

$$f(a, b, c) = \frac{1}{4}\sqrt{(a+b+c)(a+b-c)(a+c-b)(b+c-a)}.$$

Show that

$$f(a, b, c) = f(a, c, b) = f(b, a, c) = f(b, c, a) = f(c, a, b) = f(c, b, a)$$

The function $f$ is an example of what is called a *symmetric function*. Why do you think that mathematicians use the adjective *symmetric* to describe $f$?

Why does this make sense geometrically?

4.   What is the value of

$$\frac{f(5a,\ 5b,\ 5c)}{f(a,\ b,\ c)}\ ?$$

Why does this make sense geometrically?

These kinds of problems and questions illustrate that Heron's formula almost has to be what it is—its algebraic form mirrors the necessary geometric constraints. More precisely, if $f$ is an algebraic function of three variables that produces the area of a triangle from the lengths of its sides,

- $f(a, b, c)$ must be symmetric in $a$, $b$, and $c$, since the area of a triangle does not depend on the order in which the sides are labeled.

- $f(a, b, c)$ must vanish if $a = b = c = 0$ or if any two of $a$, $b$, and $c$ add to the third, since the area of such degenerate triangles is 0.

- If $k \geq 0$, $f(ka, kb, kc)$ must be $k^2 \cdot f(a, b, c)$, since the area of a triangle scales as the square of a scale factor that is applied to its sides.

- $f(a, b, c)$ must fail to be a real number if the sum of any two of $a$, $b$, and $c$ is less than the third, since such numbers cannot be lengths of the sides of a triangle.

- $f(a, a, a)$ must be the same as $\dfrac{a^2}{4}\sqrt{3}$, since such a triangle is equilateral, and the area of an equilateral triangle is given by the second expression.

There are other constraints like this. A fruitful classroom discussion can focus on a careful look at Heron's formula and how it contains "hidden meaning." Students can generate facts about triangles and their areas and then explore how these facts are found in the algebraic formula. Students with more technical algebraic skill can be asked to *derive* Heron's formula. A good preparation for an algebraic proof is to carry out the calculation with numbers, delaying the evaluation until the form of the result is apparent, and concentrating on the rhythm of the calculations.

For example, suppose that a triangle has sides of length 13, 14, and 15. Draw an altitude to one side of the triangle, calling its length $h$. as illustrated in figure 1.12.

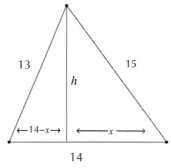

Fig. 1.12. A triangle with side lengths 13, 14, and 15

Find $h$ by using the Pythagorean theorem twice:

$$x^2 + h^2 = 225$$

and

$$(14 - x)^2 + h^2 = 169$$

Expand the second equation:

$$196 - 28x + x^2 + h^2 = 169$$

Several directions are possible from this point. *Focus in High School Mathematics: Reasoning and Sense Making* calls for "reasoned solving"; one example might be to replace $x^2 + h^2$ in the expanded equation by 225:

$$196 - 28x + 225 = 169$$

This is a linear equation in $x$; the solution is $x = 9$. This solution implies that $h = 12$, so the area of the triangle is $\frac{1}{2}(14 \cdot 12) = 84$. It would be worthwhile (especially before considering the general derivation) to determine where the height 12 comes from, expressing it in terms of 13, 14, and 15, and conducting a careful analysis of the calculations that yielded 12.

Nothing is special about 13, 14, and 15 in this situation. The same reasoning can be carried out in general. Suppose that the triangle is labeled like the one in figure 1.13, where $h$ is the altitude to side $c$.

> Stepping back to reason where the numbers in the expanded equation come from, notice that $196 = 14^2$, $169 = 13^2$, and $225 = 15^2$. What about the 28? Observe that it is twice the length of the base.

> A triangle like the one under discussion, with integer side lengths and integer area, is called a *Heron triangle*. Much has been written about Heron triangles and how to generate them. They are perfect for this activity: students can concentrate on the steps in the calculation that can be generalized with algebra without getting bogged down in the arithmetic calculations that are specific to the particular numerical values.

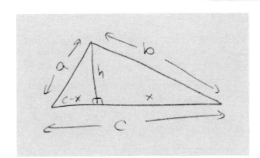

Fig. 1.13. A triangle with sides $a$, $b$, and $c$, with $h$ as the altitude to side $c$

The basic idea is to find an expression for $h$ in terms of $a$, $b$, and $c$. As in the numerical examples, we can apply the Pythagorean theorem to the two right triangles in the figure:

$$h^2 + x^2 = b^2$$
$$h^2 + (c - x)^2 = a^2$$

We can expand the second equation to obtain

$$h^2 + c^2 - 2cx + x^2 = a^2.$$

Then we can replace $h^2 + x^2$ by $b^2$ and solve for $x$:

$$x = \frac{c^2 + b^2 - a^2}{2c}$$

In considering the relation $x = \frac{c^2 + b^2 - a^2}{2c}$, look back at the numerical example. Note that $5 = \frac{14^2 + 13^2 - 15^2}{2 \cdot 14}$. The (mindful) manipulations from this point on are not explicit in the numerical example as we presented it. If you follow the same path as we did, you might go back after this derivation and retrace the steps in the numerical example, showing where each of these factorizations and simplifications is implicit.

The relation $x = \frac{c^2 + b^2 - a^2}{2c}$ and the relation $h^2 + x^2 = b^2$ allow us to find $h^2$:

$$h^2 = b^2 - x^2$$

$$= b^2 - \left( \frac{c^2 + b^2 - a^2}{2c} \right)^2$$

By simplifying, we obtain

$$h^2 = \frac{1}{4c^2} \left( 4b^2c^2 - \left( c^2 + b^2 - a^2 \right)^2 \right).$$

The expression inside the parentheses is a difference of squares, so we have

$$h^2 = \frac{1}{4c^2} \left( 2bc - \left( c^2 + b^2 - a^2 \right) \right) \left( 2bc + \left( c^2 + b^2 - a^2 \right) \right)$$

$$= \frac{1}{4c^2} \left( a^2 - \left( c^2 - 2bc + b^2 \right) \right) \left( \left( c^2 + 2bc + b^2 \right) - a^2 \right)$$

$$= \frac{1}{4c^2} \left( a^2 - (c - b)^2 \right) \left( (c + b)^2 - a^2 \right).$$

If the class has gotten this far, students can finish the factorization to show that

$$h^2 = \frac{1}{4c^2} \ (a - c + b)(a + c - b)(c + b - a)(c + b + a).$$

This leads to Heron's formula.

Students could perhaps write up this derivation of Heron's formula as a report and publish it in the school paper.

## Extending the idea

A lovely generalization of Heron's formula to quadrilaterals is within reach of students who know some trigonometry. A special case of the generalization deals with cyclic quadrilaterals—that is, quadrilaterals whose vertices lie on a circle. If the side lengths of a cyclic quadrilateral are $a$, $b$, $c$, and $d$, and if its area is $A$, then

$$A^2 = \frac{1}{16}\,(a + b + c - d)(a + b + d - c)(b + c + d - a)(a + c + d - b).$$

This relation is Brahmagupta's formula, named after a brilliant Indian mathematician who worked around 620 CE. Trigonometry students could prove Brahmagupta's formula, but even without a proof, students can investigate how the formula makes sense and dovetails with Heron's formula. For example, what happens to the formula *and* to the quadrilateral as $d$ approaches 0?

The more general Brahmagupta's formula that works for all quadrilaterals is

$$A^2 = \frac{1}{16}\,(a + b + c - d)(a + b + d - c)(b + c + d - a)(a + c + d - b) - abcd \, \cos^2\theta,$$

where $\theta$ is half the sum of the measures of two opposite angles in the quadrilateral. Consider a further opportunity for sense making: When is the "error" term $abcd \, \cos^2\theta$ equal to 0? What does this say about the geometry?

## Conclusion

This chapter has presented a few examples of ways that algebra and geometry "talk to each other." Geometry and algebra can be developed together in many other situations in high school mathematics in ways that can enhance students' understanding of both disciplines. Examples of potentially fruitful topics include the following:

- *The Pythagorean theorem.* Many proofs of the Pythagorean theorem lend themselves to algebra-geometry interpretations. For example, you might give your students four congruent right triangles and ask them to arrange the triangles so that the eight legs form an outline of a square. After some experimentation, students often produce something that looks like figure 1.14. Again, as for the proofs about area earlier in the chapter, the students need to verify certain facts: Do the two legs on each edge really form a straight line? Is the figure really a square?

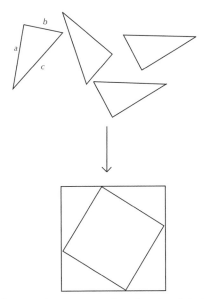

Fig. 1.14. Four congruent right triangles rearranged so that the eight legs form the outline of a square

Algebraically, the area of the large square is $(a + b)^2$, the area of each triangle is $\frac{1}{2} ab$, and the area of the inner square is $c^2$, so

$$(a + b)^2 = 4\left(\frac{1}{2} ab\right) + c^2,$$

and the Pythagorean theorem follows.

• *Tangent circles.* Suppose that you have two circles of radii $a$ and $b$, with each circle tangent to a line. Fit in a third circle of radius c so that the circle is tangent to both of the original circles and to the line (see fig. 1.15).

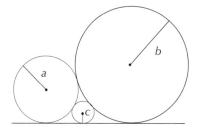

Fig. 1.15. Three circles that are mutually tangent, and tangent to the same line

Show that

$$2\left(\left(\frac{1}{a}\right)^2 + \left(\frac{1}{b}\right)^2 + \left(\frac{1}{c}\right)^2\right) = \left(\left(\frac{1}{a}\right) + \left(\frac{1}{b}\right) + \left(\frac{1}{c}\right)\right)^2 .$$

The reciprocal of a circle's radius is its *curvature.*

Are there other circles that are tangent to all three circles?

A related problem concerns *four* mutually tangent circles (see fig. 1.16).

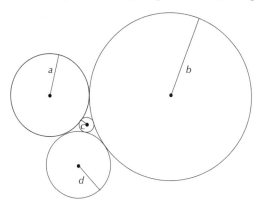

Fig. 1.16. Four mutually tangent circles

A challenge is to show that

$$2\left(\left(\frac{1}{a}\right)^2+\left(\frac{1}{b}\right)^2+\left(\frac{1}{c}\right)^2+\left(\frac{1}{d}\right)^2\right)=\left(\left(\frac{1}{a}\right)+\left(\frac{1}{b}\right)+\left(\frac{1}{c}\right)+\left(\frac{1}{d}\right)\right)^2.$$

Investigating—algebraically and geometrically—what happens as $d$ gets very large is another way to connect algebra and geometry.

- *Cyclic quadrilaterals.* Many investigations into quadrilaterals inscribed in circles provide arenas for purposeful algebraic calculations. Roger Howe of Yale University provided the authors with the following exploration involving figure 1.17.

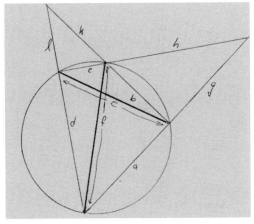

Fig. 1.17. A cyclic quadrilateral with side lengths $a$, $b$, $c$, and $d$ and diagonals of lengths $e$ and $f$

This figure can be used to verify the following proportions:

$$\frac{a}{c}=\frac{d+l}{k}=\frac{b+k}{l}$$

$$\frac{g}{h}=\frac{e}{f}=\frac{c+h}{a+g}$$

$$\frac{d}{b}=\frac{a+g}{h}=\frac{c+h}{g}$$

$$\frac{k}{l}=\frac{f}{e}=\frac{d+l}{b+k}$$

Using these and some "mindful manipulations," one can derive many interesting theorems, including Brahmagupta's formula and Ptolemy's theorem, which states that $ef = ac + bd$, where $a$, $b$, $c$, $d$, $e$, and $f$ are the lengths labeled in figure 1.17. Details can be found in Durell and Robson (2003).

# Chapter 2

# Building Equations and Functions

*Focus in High School Mathematics: Reasoning and Sense Making* (NCTM 2009) stresses the importance of reasoning with algebraic symbols, equations, and functions. These skills are precisely the ones that cause students so much difficulty in the transition from arithmetic to algebra. Indeed, teachers' assessments of the areas that cause beginning students to struggle are overwhelmingly uniform. The following areas often present challenges:

1. Expressing generality with algebraic notation, including function notation

2. Reasoning about slope, graphing lines, and finding equations of lines

3. Building and using algebraic functions

4. Setting up the appropriate equations to solve word problems

At first glance this list looks like a collection of disparate topics. Yet, looking underneath the topics and considering the kind of reasoning that would help students master them reveals a remarkable similarity. *A key component of all of these topics is the reasoning habit of seeking and expressing regularity in repeated calculations.*

This habit manifests itself when one is performing the same calculation over and over and begins to notice the "rhythm" in the operations. Articulating this regularity leads to a generic algorithm, which is typically expressed with algebraic symbols and can be applied to any instance and transformed to reveal additional meaning, often leading to a solution of the problem at hand.

This chapter explores how this habit can be used to bring coherence to three topics in the high school curriculum:

> The habit of seeking and expressing regularity in repeated calculations runs throughout the specific components that *Focus in High School Mathematics: Reasoning and Sense Making* identifies in the reasoning habits that it describes. For example, those components include—
> - identifying relevant mathematical concepts, procedures, or representations;
> - seeking patterns and relationships;
> - looking for hidden structure;
> - making purposeful use of procedures;
> - organizing the solution, including calculations.

1. Building equations to model situations

2. Finding lines of best fit

3. Calculating monthly loan payments

# From Calculations to Equations

*Focus in High School Mathematics: Reasoning and Sense Making* calls for *reasoned solving* of equations—seeing steps in the solution of an equation as logical deductions. However, before equations can be solved, they have to be constructed, by using what that publication calls the meaningful use of symbols. Teachers report that many students, even students who are quite skillful in solving linear and quadratic equations, have a very hard time building equations that model particular situations.

> Teachers of algebra typically comment, "My students can solve the equations; setting them up is the hard part."

For example, consider how hard it is for students to set up an equation that they can use to solve an algebra word problem. Reasons for their difficulties typically include the reading levels and the unfamiliar contexts of such problems. Still, there has to be more to students' difficulties than these surface features. Consider, for example, the following two problems:

**Problem 1:** The driving distance from Boston to Chicago is 990 miles. Rico drives from Boston to Chicago at an average speed of 50 mph and returns at an average speed of 60 mph. For how many hours is Rico on the road?

**Problem 2:** Rico drives from Boston to Chicago at an average speed of 50 mph and returns at an average speed of 60 mph. Rico is on the road for 36 hours. What is the driving distance from Boston to Chicago?

The problems have identical reading levels and context. But teachers report that many students who can solve problem 1 are baffled by problem 2. A significant body of research can help to explain this phenomenon (Bransford, Brown, and Cocking 1999; Breidenbach et al. 1992; Cuoco 1995; Piaget 1972; Sfard 1991; Sfard and Linchevski 1994; Slavit 1997).

> Problems 1 and 2 and others like them make no pretense of being rooted in reality. Indeed, their puzzle-like quality makes them ideal vehicles for developing the reasoning habits under consideration.

Problem 1 can be solved with isolated calculations, as shown in figure 2.1. However, problem 2 requires that the student encapsulate these isolated individual calculations into a coherent *process*—an algorithm that calculates the time on the road from the distance traveled—so that they can invert the algorithm (reasoned solving again) to come up with a distance that will produce 36 hours.

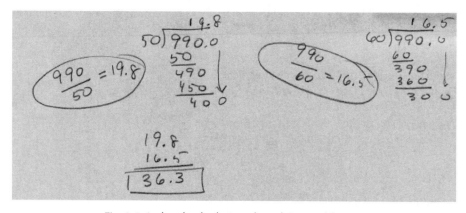

Fig. 2.1. Isolated calculations for solving problem 1

In this situation, the reasoning habit of "expressing the rhythm" in a calculation can be of great use to them. The basic idea is for them to guess at an answer to problem 2 and check their guess as if they were working on problem 1, *keeping track of their steps.* The purpose of the students'

Students who solve this problem with the aid of a calculator typically hit the "=" key very often.

guess is not to stumble on (or to approximate) the correct answer; rather, it is to help them construct a "checking algorithm" that will work for any guess. So, students can make several guesses until they are able to express their checking algorithm in algebraic symbols. The following example shows how a student might approach this problem; figure 2.1 shows the student's calculations.

*Student:*   I began by guessing that the distance is 1000 miles. I then divided 1000 by 50 and 1000 by 60. Then I added the answers together to see if I got 36. I didn't, so I made another guess—950 miles. Let's see: 950 divided by 50 plus 950 divided by 60. Is that 36? No, but a general method is evolving that might allow me to check *any* guess. My guess-checker is

$$\frac{\text{guess}}{50} + \frac{\text{guess}}{60} \stackrel{?}{=} 36.$$

So my *equation* is

$$\frac{\text{guess}}{50} + \frac{\text{guess}}{60} = 36,$$

or, letting $x$ stand for the unknown correct guess,

$$\frac{x}{50} + \frac{x}{60} = 36.$$

## In the classroom

In the following vignette, two teachers sort out the difference between the solution method described for problem 2 and traditional guess-and-check. Mr. Thomas Gradgrind and Ms. Maria Agnesi are talking about their algebra classes. Tom is sharing his concerns about a lesson involving the relationships among distance, rate, and time.

"Guess-and-check" has long been a popular method for finding or approximating solutions to all kinds of problems. What we present here isn't quite the same—the guesses are just *scaffolds* to help students build equations. The real goal is to build a generic "guess checker"—the equation that can be solved to produce the exact solution.

*Tom:*   Maria, I just don't know what to do. Right now in my class we are working on distance-rate problems. We had already talked about the relationship among distance, rate, and time. I then gave students a problem like this: "The driving distance from Boston to Chicago is 990 miles. Rico drives from Boston to Chicago at an average speed of 50 mph and returns at an average speed of 60 mph. How many hours was Rico on the road?" Almost every student was able to come up with the correct solution.

*Maria:*   How do you know students understood what they were doing?

| | |
|---|---|
| *Tom:* | When I walked around to see what students were doing, I saw that they were dividing the one-way distance by each respective speed and then adding both times to get the total hours. I asked students to explain why they were dividing, and they were able to talk about $d = rt$. |
| *Maria:* | So what exactly is your concern? |
| *Tom:* | After asking students to determine the total time for the problem, I switched the problem a bit. I gave students the same speeds as before but told them this time that Rico was making a round trip from Fort Lauderdale, Florida, to Reston, Virginia. I asked them to figure out the one-way distance between the two cities if the total driving time was 38.5 hours. They didn't even know how to begin the problem. So I ended up just telling them how to set up the equation to solve the problem. |
| *Maria:* | What understanding do you think your students have about the problem? |
| *Tom:* | None. I gave them a formula of sorts that can help them solve these types of problems. What else was I supposed to do? |
| *Maria:* | This is a great opportunity to help students develop as problem solvers while at the same time giving them a chance to make meaning out of algebraic symbols. Let me show you what I mean. Given two different rates, one each for the trip out and back, your students were able to determine the total trip time. Well, have them use this method to help solve the second problem. |
| *Tom:* | I'm not sure that I follow. Students didn't set up an equation initially, and they clearly couldn't set up an equation for the second problem. |
| *Maria:* | Suggest to students that they "guess" a distance and use it to check if they are correct by calculating if they get the same total driving time. |
| *Tom:* | But how does guessing help them? I don't want them to keep guessing and checking. It's not efficient, and they may never get the right answer. |
| *Maria:* | The "guessing" is just the means for them to develop an algorithm. Have students keep track of the steps they are using to check their guess. Here, let's try one. Begin with a guess of 500 miles and conjecture what students will do. |
| *Tom:* | They will divide 500 by 50 and then divide 500 by 60 and add them together to get the total time—just as they did for the initial problem. |
| *Maria:* | Suggest they try another number for the distance between the cities, like maybe 800 miles. What will they do? |
| *Tom:* | The same thing as before. They will divide 800 by 50 and then divide 800 by 60 and add them together. Oh, I see what you're getting at. After a couple of times, students can begin to see a pattern. I can coach them to come up with a type of verbal description, like |

$$\frac{\text{Miles between cities}}{50 \text{ mph}} + \frac{\text{Miles between cities}}{60 \text{ mph}} = 38.5 \text{ hours}.$$

*Maria:*        Now students can simply replace "miles between cities" with $x$ and they have an equation where the variables and equation make sense to them. They have also developed a method that will come in very handy in the future for setting up equations.

This habit of trying numerical examples until the structure of an algorithm becomes clear captures a very common process that is a useful tool throughout algebra: we carry out several concrete examples of a process that we don't quite "have in our heads" to find regularity and build a generic algorithm that describes every instance of the calculation. As another example, let's look at how this same reasoning can be used to find equations of lines and other curves.

## Equations of lines and other curves

Suppose that a student who is new to algebra and comes to it with no formulas is asked to find the equation of the vertical line $l$ that passes through the point with coordinates $(5, 4)$. Students can draw the line, and, just as in the

> Roger Howe (forthcoming) makes a careful analysis of word problems, showing how arithmetic and algebraic approaches can be developed and used in tandem.

word problem example, they can guess points and check to see if they are on $l$. For example, trying some points, like $(5, 1)$, $(3, 4)$, $(2, 2)$, and $(5, 17)$, leads to a generic guess-checker: *To see if a point is on $l$, you check that its $x$-coordinate is 5.* This leads to the guess-checker $x \overset{?}{=} 5$ and the equation $x = 5$.

> To be completely rigorous, students should check that a point is on $l$ if and only if its $x$-coordinate is 5. The equation $x = 5$ is often referred to as a *point tester* for $l$.

This method works well for vertical and horizontal lines, and even for special lines like the one that bisects the first and third quadrant. But what about lines for which there is no simple guess-checker? The idea is to find a geometric characterization of such a line and then to develop a guess-checker based on that characterization. One such characterization uses *slope*.

In first-year algebra, students study slope, and one fact about slope that often comes up is that three points on the coordinate plane but not all on the same vertical line are collinear if and only if the slope between any two of them is the same. Figure 2.2 shows three points that satisfy this condition and three points that do not.

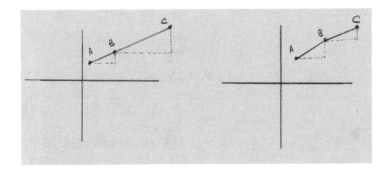

Fig. 2.2. On the left, points *A*, *B*, and *C* are collinear. On the right, they are not.

In the figure, if we let $m(A, B)$ denote the slope between $A$ and $B$ (calculated as change in $y$ divided by change in $x$), then the collinearity condition can be stated like this:

Three points $A$, $B$, and $C$ that do not all lie on the same vertical line are collinear if and only if $m(A, B) = m(B, C)$.

To prove this characterization of collinearity, one needs some facts about similar triangles. In figure 2.2, the two triangles on the left are similar; the two triangles on the right are not.

This criterion for collinearity can be used to find the equation of a line between two points. Suppose, for example, that students are asked to find an equation for $\overrightarrow{AB}$ if $A$ is the point $(5, 1)$ and $B$ is the point $(-3, 6)$. Imagine again that they have no knowledge of $y = mx + b$ or related formalisms. They can reason as follows: The slope between $A$ and $B$ is $-\frac{5}{8}$, and we can guess at some points $C$ and check to see whether or not $C$ is collinear with $A$ and $B$ by checking slopes:

| $C$ | $m(B, C)$ | $C$ on $\overrightarrow{AB}$ ? |
|---|---|---|
| $(1, 3)$ | $-\dfrac{3}{4}$ | no |
| $(7, 0)$ | $-\dfrac{3}{5}$ | no |
| $(13, -4)$ | $-\dfrac{5}{8}$ | yes |
| $(-6, 7)$ | $-\dfrac{1}{3}$ | no |

How might the students check a generic guess, say $C(x, y)$? They could calculate the slope between $C(x, y)$ and $B(-3, 6)$, and see if the slope is $-\frac{5}{8}$. The guess-checker is $m(B, C) \stackrel{?}{=} -\frac{5}{8}$, or

$$\frac{y-6}{x+3} \stackrel{?}{=} -\frac{5}{8}.$$

So an equation of $\overrightarrow{AB}$ is

$$\frac{y-6}{x+3} = -\frac{5}{8}.$$

From here, the students can simplify the equation to get it into a more standard form.

This outline glosses over some important details that would need classroom discussion. For example, the special case when $x = -3$ needs attention, and students should check this result against the result obtained when one checks the slope from $C$ to $A$ instead of from $C$ to $B$.

It is certainly true that algebra students need to become fluent in understanding the correspondence between linear equations and their graphs. In many applications, they will need to be able to read the slope and $y$-intercept of a line from its equation, and given these features, they will need to be able to draw a line.

So, why not jump directly to the development of these skills without the guess-checking activities? A number of reasons support starting with an approach like the one outlined here:

> Students frequently think that $y = 3x + 4$ is a "code" that means "put a point at (0, 4), then go over 1 and up 3, put a point there, and then draw a line between these two points."

1. Several research studies (Greenes et. al. 2007; Goldenberg 1988, 1991) show that students who can fluently graph equations like $y = 3x + 4$ often can't use the equation to see if a given point is on the graph. Building equations from the repeated testing of numerical examples reinforces the "Cartesian connection" that a point is on the graph of an equation if and only if its coordinates satisfy the equation.

2. This same reasoning habit can be applied to other equations and their graphs. For example, to find an equation for the circle with center $C(3, 7)$ and radius 5, students who are used to thinking this way might ask, "How can I check to see if a given point $P$ is on the circle?" They might then follow up this question by asking, "Is the distance from $P$ to $C$ equal to 5?" Students equipped with the Pythagorean theorem would be able to write down the equation from this characterization long before learning about the formula $(x - h)^2 + (y - k)^2 = r^2$.

3. The very act of articulating a guess-checking algorithm in a way that can be formulated with algebraic symbols is a skill that will serve students well throughout mathematics and related fields.

*Automaticity* in graphing is very important. However, jumping directly to the automatic applications of methods like using "$y = mx + b$" can disconnect students' skill in graphing equations from the underlying meaning that connects equations and their graphs.

> Articulating a guess-checking algorithm as described will serve students well in dealing with algebra word problems and area formulas. Eventually, students should be able to go from a problem directly to an equation or function that models the problem's situation, but jumping directly to "problems by type" or rules like "let x= ..." or $A = \frac{1}{2} (b_1 + b_2)h$ can disconnect the symbols from their meaning for students.

## Fitting Lines to Data

Imagine a class in which students have developed automaticity with the connection between lines and their equations. One application of this set of skills is to provide some insight into the sometimes-mysterious calculator button that calculates the line of best fit for a set of data. After students have had appropriate informal experiences with data trends, many high school curricula give a definition of a best-fit line in a manner such as the following:

For a set of points $\{(x_i, y_i)\}_{i=1}^n$, the line of best fit minimizes the sum of the squares of the deviations in $y$-values. In other words, it is a line with equation $y = ax + b$ so that the sum

$$\sum_{i=1}^{n} \left( y_i - (ax_i + b) \right)^2$$

is as small as possible.

> Notice that $a$ and $b$ are the variables here. Think of the set of all possible lines dancing through the data, each one with its own "badness," or lack of fit (its particular sum of squares of deviations in $y$-heights from the data points). The use of dynamic geometry software can make this image precise (Cuoco and Goldenberg 1996).

The actual derivation of the $a$ and $b$ that minimize this sum is usually left for linear algebra or calculus. However, a little knowledge of quadratic functions (and how to minimize them), along with the habit of abstracting from calculations, can take students quite a bit further.

*Best* in the case of "line of best fit" refers to the line that minimizes the sum of the squares of the deviations in *y*-coordinates. Finding the minimum with respect to this metric, although mathematically possible, may not be useful in a given situation. See Good and Hardin (2006).

Let's start with a very simple example. Suppose that the following table shows our data:

| Input | Output |
|-------|--------|
| 1 | 3 |
| 2 | 4.5 |
| 3 | 8.1 |
| 4 | 8 |

A popular experiment that uses dynamic geometry software allows students to plot the data, construct a moveable line through them, and (dynamically) calculate the sum of the squares of the differences between the *y*-values of the data and the corresponding *y*-values on the line. Figure 2.3 shows the differences for a given line.

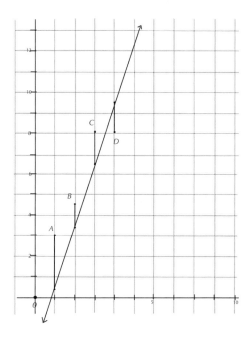

Fig. 2.3. The differences of the *y*-values of data points and the corresponding *y*-values for a given line

Playing with the experiment develops an awareness of two degrees of freedom: the line's slope and its height above the *x*-axis (its *y*-intercept). In such situations, we can hold one variable fixed and vary the other. Suppose that we fix the slope at—say, 3—and ask a more specific question:

Of all possible lines of slope 3, which one best fits the data?

Such lines have equations $y = 3x + b$ for varying $b$, so we can make a table as in figure 2.4. Note that the error is the difference between the actual output and value predicted by the linear equation.

Data vs. Predictions Given by $y = 3x + b$

| Input | Output | Predicted | Error |
|-------|--------|-----------|-------|
| 1 | 3 | $3 \cdot 1 + b = 3 + b$ | $3 - (3 \cdot 1 + b) = -b$ |
| 2 | 4.5 | $3 \cdot 2 + b = 6 + b$ | $4.5 - (3 \cdot 2 + b) = -1.5 - b$ |
| 3 | 8.1 | $3 \cdot 3 + b = 9 + b$ | $8.1 - (3 \cdot 3 + b) = -.9 - b$ |
| 4 | 8 | $3 \cdot 4 + b = 12 + b$ | $8 - (3 \cdot 4 + b) = -4 - b$ |

Fig. 2.4. Data vs. predictions given by $y = 3x + b$

So, we want to minimize

$$(-b)^2 + (-1.5 - b)^2 + (-.9 - b)^2 + (-4 - b)^2.$$

This expression is *quadratic* in b, and by the end of first-year algebra, students know how to minimize a quadratic. The above expression simplifies (by hand or computer algebra system [CAS]) to

$$4b^2 + 12.8b + 19.06.$$

Hence, the minimum value occurs when

$$b = \frac{-12.8}{2 \cdot 4} = -1.6.$$

Thus, of all lines with slope 3, the one that is the best fit for the data has equation

$$y = 3x - 1.6.$$

Students can play the same game with different slopes, and a certain rhythm to the calculations develops, allowing them to construct a table like the one in figure 2.5. The "abstracting regularity" habit plays out a little differently in this instance: here it is used to *generate* a table of results from which a pattern emerges.

Best-Fit Line: $y = ax + b$

| Slope (a) | "Badness" | Minimizing value of b |
|:---:|:---:|:---:|
| 0 | $158.86 - 47.2b + 4b^2$ | 5.9 |
| 1 | $52.26 - 27.2b + 4b^2$ | 3.4 |
| 2 | $5.66 - 7.2b + 4b^2$ | .9 |
| 3 | $19.06 + 12.8b + 4b^2$ | −1.6 |
| 4 | ... | −4.1 |

Fig. 2.5. Slopes and corresponding minimized b-values for best-fit line

Students can investigate how $b$ grows with $a$ by looking at successive differences; see the table in figure 2.6. In this table, the $\Delta$ column gives the difference between the best $b$ at row $n$ and the best $b$ at row $n + 1$ (for $n = 0, ..., 3$).

Change in b for Best-Fit Line: $y = ax + b$

| Slope (a) | Equation of best-fit line | y-intercept (b) | Δ |
|:---:|:---:|:---:|:---:|
| 0 | $y = 5.9$ | 5.9 | −2.5 |
| 1 | $y = x + 3.4$ | 3.4 | −2.5 |
| 2 | $y = 2x + .9$ | .9 | −2.5 |
| 3 | $y = 3x - 1.6$ | −1.6 | −2.5 |
| 4 | $y = 4x - 4.1$ | −4.1 | |

Fig. 2.6. Change in b for best-fit line

The $y$-intercept seems to depend *linearly* on the slope. It seems as if the best-fit line of slope $a$ has equation $y = ax + b$, where

$$b = 5.9 - 2.5a.$$

The graphs of all of these lines show what this condition implies; see figure 2.7.

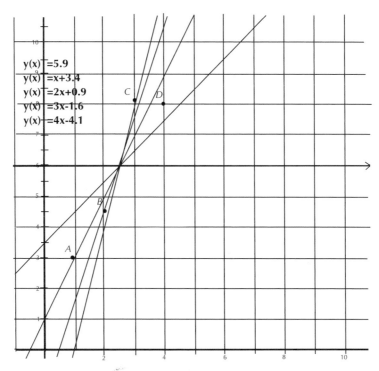

Fig. 2.7. Graphs of best-fit lines from figure 2.6

The lines seem to be *concurrent*, and the point of concurrency is (2.5, 5.9). This point is the *centroid* of the data—the point whose $x$ and $y$ coordinates are the averages of the $x$- and $y$-coordinates of the data. (The centroid is often denoted by $(\bar{x}, \bar{y})$ and is sometimes called the *balance point* of the data.)

That the line of best fit with a fixed slope passes through the centroid of the data is a fact that advanced algebra students can prove, and again, the habit of seeking regularity in calculations can help them build up to the proof. Yet, there are many heuristic arguments for assuming this fact even before students are able to prove it. By making this fact an assumption (an unproved lemma, for the time being), students can concentrate on finding, of all lines that pass through

> The concurrency of the lines can be established by finding the intersection of any two lines and then verifying that this intersection point is on all the others.

the centroid, the *slope* of the line that best approximates the data. This work again involves minimizing a quadratic—this time a quadratic in the slope $a$.

Finally, having assumed (or having proved) this fact about the centroid, students can employ the reasoning habit of seeking regularity in repeated calculations and can gradually build up an algorithm that will produce the best-fit line for *any* set of data. This algorithm can be further encapsulated in a formula for the line of best fit—a *formula* that is usually derived in calculus—by using only methods from high school algebra (Education Development Center 2009).

# Recursive Thinking

When using functions to describe situations, we often find it more natural to use recursive definitions rather than the typical closed-form expressions like polynomials or rational expressions. *Focus in High School Mathematics: Reasoning and Sense Making* includes two examples—example 11, Take As Directed (pp. 45–48), and example 12, Money Matters (pp. 49–51)—that present cases where recursively defined functions closely model the situations at hand.

Only a few decades ago, the ability to model situations with recursively defined functions was considered a very difficult and advanced skill. In the latter half of the twentieth century, however, electronic computational environments that supported recursively defined functions—Logo, Pascal, and spreadsheets, for example—were instrumental in making recursion much more tractable for precollege students. Careful use of the descendents of these tools—especially those that have a syntax that is a good approximation of ordinary mathematical notation—remains extremely useful. Although many students now find recursive thinking quite natural, many also find it very hard to express recursion mathematically.

Many Web sites calculate the monthly payment on a car listed at a given price and purchased with a given amount of money down and a loan at a given rate, extending over a given number of months. See "Car Financing Advice" at http://www.cars.com.

This section builds on example 12, Money Matters, and takes a closer look at the problem of determining the monthly payment on a loan, highlighting the value added by the use of a modern computer algebra environment. Our example will also show how the habit of abstracting regularity from calculations can help students build recursively defined functions.

Imagine an activity in which high school students investigate the question of how a bank figures out the monthly payment on a loan. Many car dealerships have Web sites that allow customers to enter the price of a listed car, an interest rate, and the number of months in which they want to pay off the loan. The program then tells prospective buyers the monthly payment. Suppose that students explore one such Web site that lists a car for $27,355. The Web site reports that with a down payment of $1,000, a term of 36 months, and a 5% interest rate, the monthly payment would be $789.88. Students might wonder how that amount was calculated.

Such functions are built into many calculators, but actually figuring out the algorithm leads to some nice algebra. One way to think about this problem is to develop an algorithm that calculates the balance on the loan at the end of each month as a function of the monthly payment. In the example above, we want to find the monthly payment that makes the balance at the end of 36 months equal to 0. Let $b(n, m)$ represent the balance at the end of $n$ months with a monthly payment of $m$ dollars, and let's start with a simple example: suppose that there is no interest on the loan. Then the balance that a car buyer owes at the end of any month is just the balance owed at the start of the month minus the monthly payment. Our model would look like this:

$$b(n, m) = \begin{cases} 26355 & \text{if } n = 0 \\ b(n-1, m) - m & \text{if } n > 0 \end{cases}$$

The model tells us that $b(0, m)$—what the car buyer owes when driving out of the dealership—is in this case $26,355. After that, the balance at the end of any month is the balance at the start of the month minus the monthly payment. In figure 2.8, the screenshot using TI-Nspire software shows the faithfulness of the syntax of the computational model to the mathematical notation.

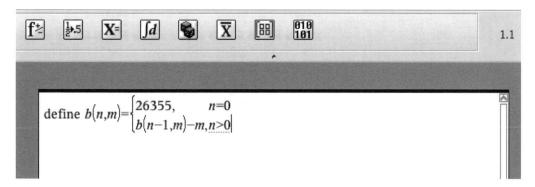

Fig. 2.8. With TI-Nspire software, the syntax of the computational model is faithful to the mathematical notation.

Students can experiment with this model, tabulating it in a spreadsheet, for example. Without interest, the model simply subtracts $m$ dollars from the previous balance—the "regularity" in the calculation is that the balance decreases by $m$ as $n$ increases by 1. Repeatedly subtracting $m$ 36 times amounts to subtracting $36m$. Hence, the balance at the end of 36 months is given by

$$26355 - 36m.$$

Setting this expression equal to 0 and solving for $m$ gives the monthly payment with no interest.

However, most cases involve a nonzero interest rate. The algorithm that most lenders use in this case is the following:

What a borrower owes at the end of the month is what the borrower owed at the start of the month, plus $\frac{1}{12}$ th of the yearly interest on that amount, minus the monthly payment.

Working out the payments for a couple of months by hand until the rhythm of the calculations emerges leads to a modification of our original model in the case of a 5% annual interest rate:

$$b(n, m) = \begin{cases} 26355 & \text{if } n = 0 \\ b(n-1, m) + \dfrac{.05}{12} b(n-1, m) - m & \text{if } n > 0 \end{cases}$$

Figure 2.9 shows a screenshot with TI-Nspire software showing the revised model in this case.

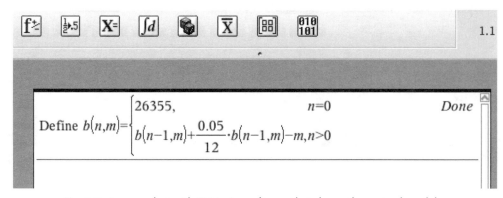

Fig. 2.9. A screenshot with TI-Nspire software that shows the revised model

Mathematically, this model is correct. But it is extremely inefficient computationally—each call to the function $b$ causes the computer to call two more instances of $b$. Algebra can come to the rescue. In particular, the distributive property permits writing the definition for $b$ in a way that will be much more efficient when implemented in a CAS:

$$b(n,\ m) = \begin{cases} 26355 & \text{if } n = 0 \\ \left(1 + \dfrac{.05}{12}\right) b(n-1,\ m) - m & \text{if } n > 0 \end{cases}$$

In figure 2.10, a screenshot with TI-Nspire software presents a more efficient model.

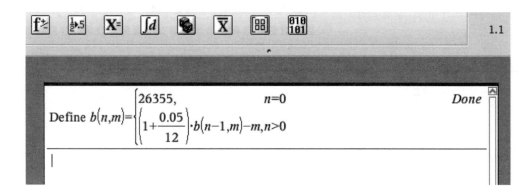

Fig. 2.10. A screenshot with TI-Nspire software that shows the more efficient model

Students now have a laboratory in which they can experiment with different monthly payments, trying to find the value of $m$ that makes $b(36, m)$ equal to 0. The screenshot in figure 2.11 illustrates how students typically obtain the monthly payment down to the penny.

This would be a good opportunity to discuss with students the reason that the first CAS model will stall even on rather small values of $n$. If they trace the process for—say, $b(15, 500)$—they will see that the needed memory grows exponentially. In a CAS, recursively defined functions can be extremely "expensive" in terms of the memory that they require for their execution.

| Define $b(n,m)=\begin{cases}26355, & n=0\\ \left(1+\dfrac{0.05}{12}\right)\cdot b(n-1,m)-m, & n>0\end{cases}$ | Done |
|---|---|
| $b(36,600)$ | 7358.6 |
| $b(36,700)$ | 3483.27 |
| $b(36,800)$ | -392.068 |
| $b(36,750)$ | 1545.6 |
| $b(36,775)$ | 576.766 |
| $b(36,785)$ | 189.232 |
| $b(36,795)$ | -198.301 |
| $b(36,790)$ | -4.5344 |
| $b(36,787)$ | 111.726 |
| $b(36,789)$ | 34.2189 |
| $b(36,789.5)$ | 14.8423 |
| $b(36,789.75)$ | 5.15393 |

Fig. 2.11. Students use approximation to find the monthly payment that will produce a balance of 0 at the end of 36 months.

Using this recursive function approach offers several advantages, even though explicit formulas and built-in calculator functions will calculate monthly payments:

1. Translating a verbal description of a process into precise mathematical language is extremely difficult for many students. This is especially true in the case of recursive processes, which many students find "easy to say" but difficult to formulate in algebraic symbols. This activity, especially when "bundled" with a computational environment that supports the kind of notation illustrated in the discussion, can help students articulate, refine, and experiment with recursive definitions.

2. Students gain a sense of mathematical power when they develop the mathematics that underlies what seem to be mysterious calculations, like the monthly payment on a loan. By using their own models, they are more able to experiment with and analyze the effects of parameters—one of the key elements identified for reasoning about functions in *Focus in High School Mathematics: Reasoning and Sense Making*.

3. Many students find this particular approach quite engaging—they love to get the monthly payment down to the penny. Because they are often quite surprised by some of the results of their experiments, some students look very diligently for underlying structures that explain what they conjecture.

There are several directions in which you might go from here with your students:

• Ask the class to find the monthly payments for cars that cost—say, $10,000, $11,000, $12,000, …, $25,000. The work might seem tedious, but the result is often a surprise to students, and the payoff can be great. A snapshot of one student's work appears in figure 2.12. (The $y$-value is the cost of the car in thousands of dollars and $y(x)$ is the monthly payment based on a fixed interest rate and term.) The surprise for many students is that the monthly payment seems to depend linearly on the cost of the car.

(a) table

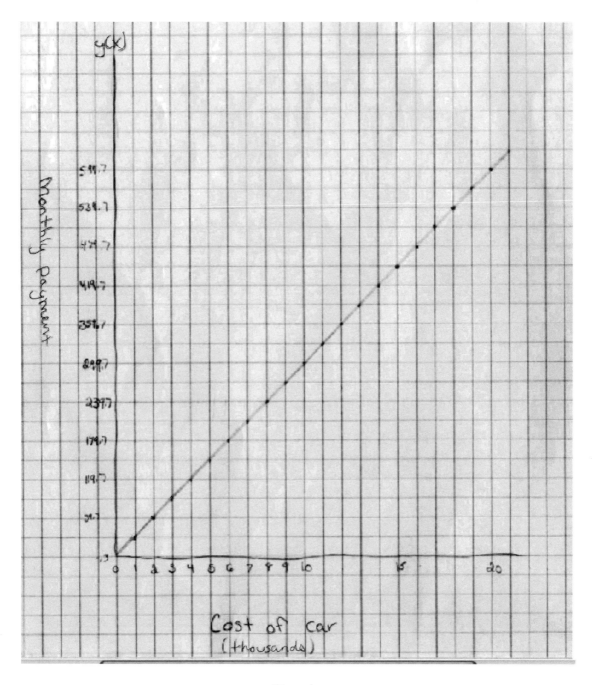

(b) graph

Fig. 2.12. A student's (a) table and (b) graph for the monthly payment on a car, depending on its cost

- Instead of varying the cost of the car, students can experiment with how the monthly payment depends on the term of the loan or on the interest rate. They might want to modify their definition of $b$ to include an interest rate that can be changed on the fly. In the model below, the interest rate is represented as a decimal:

$$b(n,\, m,\, i) = \begin{cases} 26355 & \text{if } n = 0 \\ \left(1 + \dfrac{i}{12}\right)b(n-1,\, m,\, i) - m & \text{if } n > 0 \end{cases}$$

- If students have access to a CAS, they can enter a *variable* monthly payment into their computational model and get an expression for the monthly payment at the end of 36 months, as in figure 2.13.

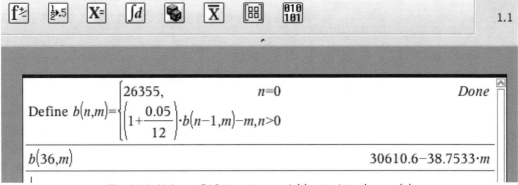

Fig. 2.13. Using a CAS to enter a variable rate into the model

This process produces another surprise: the CAS returns a linear expression. By analyzing the calculation of $b$'s output, students might try to interpret the coefficients in this expression. Working by hand or with a CAS, students can use this expression to find the exact monthly payment, as illustrated in figure 2.14. The extra precision is obtained by copying the result of the previous line rather than entering what is visible on the screen.

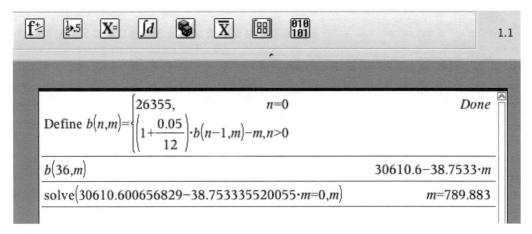

Fig. 2.14. Analyzing the outputs of $b$ enables students to determine the exact monthly payment.

## Extending the idea

An analysis of the calculations embedded in the recursive definition of function $b$ leads to an explanation of all the linearity that peeks out in these investigations. The thinking involved in the analysis provides examples of many of the reasoning habits described in *Focus in High School Mathematics: Reasoning and Sense Making*, including seeking patterns and relationships and looking for hidden structure.

Imagine a class in which students have studied geometric series and in which they have developed the habit of looking for regularity in calculations. The class might explore the following example:

Suppose that you borrow $12,000 at 5% interest to buy a car. Then you are experimenting with the function

$$b(n, m) = \begin{cases} 12000 & \text{if } n = 0 \\ \left(1 + \dfrac{.05}{12}\right) b(n-1, m) - m & \text{if } n > 0 \end{cases}.$$

Notice that

$$1 + \frac{.05}{12} = \frac{12.05}{12}.$$

Call this number $q$. So, the function now looks like

$$b(n, m) = \begin{cases} 12000 & \text{if } n = 0 \\ q \cdot b(n-1, m) - m & \text{if } n > 0 \end{cases},$$

where $q$ is a constant.

> "Call this number $q$" is an example of "chunking"—the algebraic habit of hiding complexity by naming expressions. See pages 49–52 for more discussion of this technique.

Then at the end of $n$ months, you could "unstack" the calculation as follows:

$$\begin{aligned} b(n, m) &= q \cdot b(n-1, m) - m \\ &= q\left[q \cdot b(n-2, m) - m\right] - m \\ &= q^2 \cdot b(n-2, m) - qm - m \\ &= q^2\left[q \cdot b(n-3, m) - m\right] - qm - m \\ &= q^3 \cdot b(n-3, m) - q^2 m - qm - m \\ &\vdots \\ &= q^n \cdot b(0, m) - q^{n-1} m - q^{n-2} m - \cdots - q^2 m - qm - m \\ &= 12000 \cdot q^n - m\left(q^{n-1} + q^{n-2} + \cdots + q^2 + q + 1\right) \end{aligned}$$

The last expression in parentheses is a geometric series:

$$q^{n-1} + q^{n-2} + \cdots + q^2 + q + 1 = \frac{q^n - 1}{q - 1}$$

So, you have

$$b(n,\, m) = 12000 \cdot q^n - m\left(q^{n-1} + q^{n-2} + \cdots + q^2 + q + 1\right)$$

$$= 12000 q^n - m\left(\frac{q^n - 1}{q - 1}\right).$$

Setting $b(n,\, m)$ equal to 0 gives an explicit relationship between $m$ and the cost of the car:

$$m = 12000 \frac{(q-1)q^n}{q^n - 1}$$

or, in general,

$$\text{monthly payment} = \text{cost of car} \times \frac{(q-1)q^n}{q^n - 1},$$

where $n$ is the term of the loan and

$$q = 1 + \frac{\text{interest rate}}{12}.$$

This exploration explains the linearity in the previous investigations, and it can help in finding an interpretation of the coefficients in the linear expression in figure 2.13.

## Conclusion

The examples described in the preceding sections are meant simply to be illustrative. What is important is not the topics or contexts but the reasoning habits that the contexts support. This chapter has focused on the reasoning habits that help students build equations and functions that model particular situations. Students can exercise these habits in many other arenas. One rich circle of ideas concerns the problem of fitting "simple" functions to data. This concluding section provides very brief sketches of two avenues into this subject.

1. ***Difference calculus.*** Many middle school curricula develop the idea that if the first differences are constant in a table of data in which the inputs increase regularly, the data can be modeled with a linear function. Consider the data in the table at the top of the next page. Note that the way in which the table is constructed leads to an "up and over" property; that is, every output is the sum of the entries above it and over to the right by one.

| Input | Output | Δ |
|-------|--------|---|
| 0 | 3 | 5 |
| 1 | 8 | 5 |
| 2 | 13 | 5 |
| 3 | 18 | 5 |
| 4 | 23 | 5 |
| 5 | 28 | 5 |
| 6 | 33 | |

> In the observation that the data in the table exhibit an "up and over" property, we once again see the "abstracting from calculations" reasoning habit.

Repeated applications of this up and over property show that the output for 6 is $6 \times 5 + 3$:

$$
\begin{aligned}
33 &= 28 + 5 \\
&= (23 + 5) + 5 = 23 + 2 \cdot 5 \\
&= (18 + 5) + 2 \cdot 5 = 18 + 3 \cdot 5 \\
&= (13 + 5) + 3 \cdot 5 = 13 + 4 \cdot 5 \\
&= (8 + 5) + 4 \cdot 5 = 8 + 5 \cdot 5 \\
&= (3 + 5) + 5 \cdot 5 = 3 + 6 \cdot 5
\end{aligned}
$$

Using the same reasoning, we see that the function $f$ defined by $f(n) = 5n + 3$ will fit the data in the table.

Exactly the same method (and the same reasoning habits) can be used to find a polynomial of minimal degree that fits a table where the differences are not constant. These ideas go back to Newton, and they connect to Pascal's triangle and binomial coefficients. Cuoco (2003) provides more details about fitting polynomials to tables.

2.  *Lagrange interpolation.* Difference calculus requires that the inputs of a table form an arithmetic sequence (they change by a constant amount). Often, data do not come this way.

    Suppose that we want to fit the data in the following table with a polynomial function:

| Input | Output |
|-------|--------|
| 1 | 2 |
| 3 | 74 |
| 4 | 179 |
| 5 | 354 |

We can use the following method to find the least-degree polynomial function that fits the data in the table. Suppose $f(x)$ is a function that will work. Rather than writing $f(x)$ as a combination of powers of $x$, we can conveniently write it in what seems like a much more complex way:

> Many polynomial functions fit the data in the table above. We are looking for one of them.

$$f(x) = A(x-1)(x-3)(x-4) + B(x-1)(x-3)(x-5) + C(x-1)(x-4)(x-5) + D(x-3)(x-4)(x-5),$$

where $A, B, C,$ and $D$ are numbers to be determined. The reason for choosing this form is that it makes it easy to evaluate the expression at the inputs of the table:

$$2 = f(1) = -24D \quad \text{so } D = -\frac{1}{12}$$

$$74 = f(3) = 4C \quad \text{so } C = \frac{37}{2}$$

$$179 = f(4) = -3B \quad \text{so } B = -\frac{179}{3}$$

$$354 = f(5) = 8A \quad \text{so } A = \frac{177}{4}$$

Figure 2.15 illustrates the use of a CAS to expand and simply the expression for $f(x)$ at these inputs.

$$\text{expand}\left(\frac{177}{4}\right)\cdot(x-1)\cdot(x-3)\cdot(x-4)+\frac{-179}{3}\cdot(x-1)\cdot(x-3)\cdot(x-5)+\frac{37}{2}\cdot(x-1)\cdot(x-4)\cdot(x-5)+\frac{-1}{12}\cdot(x-3)\cdot(x-4)\cdot(x-5)$$

$$3\cdot x^3 - x^2 + x - 1$$

Fig. 2.15. Using a CAS to expand and simplify the expression for $f(x)$

The calculations show that $f(x) = 3x^3 - x^2 + x - 1$ agrees with data in the table. This method is called *Lagrange interpolation*, and it has many practical and theoretical applications. Lagrange interpolation can be introduced and connected to major results in algebra 2. It's a good example of how reasoning about calculations can shed insight into what might seem to be a very difficult problem.

# Chapter 3

# Formal Algebra

Algebra, especially as it has evolved over the past two centuries, has two faces. On one hand, an algebraic expression or function is an invitation to calculate—an invitation to replace the variables with numbers and to follow the instructions given by the operations in the expression. On the other hand, such expressions are objects in their own right—elements of a formal system with its own internal logic and structure.

This "function-form" dialectic pervades school algebra. When, as illustrated in the last chapter, we build a function to calculate the monthly payment on a loan, we are thinking of the variables $m$ and $n$ as placeholders for numbers—the monthly payment and the term of the loan, respectively—and the function as a set of instructions for figuring out the balance at the end of $n$ months with a monthly payment of $m$ dollars. But in a probability class, when we use the coefficients of $(t + h)^{10}$ to figure out the theoretical distribution of heads and tails when a fair coin is tossed 10 times, we are *not* thinking of $t$ and $h$ as numbers. Rather, they are bookkeeping mechanisms—formal symbols that help us keep track of the number of ways in which we can get combinations of heads and tails.

*Focus in High School Mathematics: Reasoning and Sense Making* describes the interplay between these two faces of algebra and suggests a progression of abstraction:

> Although a long-term goal of algebraic learning is a fluid, nearly automatic facility with manipulating algebraic expressions that might seem to resemble what is often called "mindless manipulation," this ease can best be achieved by first learning to pay close attention to interpreting expressions, both at a formal level and as statements about real-world situations. At the outset, the reasons and justifications for forming and manipulating expressions should be major emphases of instruction (Kaput, Blanton, and Moreno 2008). As comfort with expressions grows, constructing and interpreting them require less and less effort and gradually become almost subconscious. The true foundation for algebraic manipulation is close attention to meaning and structure. (NCTM 2009, pp. 31–32)

Transforming polynomials as formal expressions involves thinking about the applications of the basic rules of algebra, without thinking of the variable as a placeholder for numbers. The expressions that result from transforming a polynomial formally

> Graphing calculators support the idea that an algebraic expression defines a function. Computer algebra systems can help students develop the idea that such expressions are formal objects, elements of an algebraic structure.

are equal under any substitution. So because $x^2 - 1 = (x - 1)(x + 1)$ as polynomials, we are assured without calculating that

$$3^2 - 1 = (3 - 1)(3 + 1),$$

$$(4 + \sqrt{5})^2 - 1 = (4 + \sqrt{5} - 1)(4 + \sqrt{5} + 1),$$

and even

$$(3 + 2z + z^3)^2 - 1 = (3 + 2z + z^3 - 1)(3 + 2z + z^3 + 1).$$

The "form implies function" principle tells us that a polynomial identity produces infinitely many numerical (and algebraic) identities.

When $x$ in $x^2 - 1$ is thought of in terms of the basic rules of algebra instead of as a placeholder for a number, it is often called an *indeterminate* rather than a *variable*.

This chapter presents ideas about how students might make sense of the algebra of formal expressions in the high school curriculum in the following contexts:

1. Factoring polynomials and completing the square

2. Examining combinatorial phenomena such as the distribution of sums in tosses of number cubes

3. Exploring patterns in the factors of a sequence of polynomials

It is important to keep in mind the advice of *Focus in High School Mathematics: Reasoning and Sense Making*: "For students … with insufficient experience in reasoning and sense making, taking a formal algebra course prematurely will result in missed opportunities … [and] be counterproductive unless students have the necessary prerequisite skills" (NCTM 2009, p. 103). Although form implies function, the functional perspective—where the variables are defined and derive meaning from some particular context—is a necessary prerequisite to the formal perspective.

# Mindful Manipulation

*Focus in High School Mathematics: Reasoning and Sense Making* describes "mindful manipulation" in algebra in the following way:

> Mindful manipulation includes learning algebraic manipulation as a process guided by understanding and goals (how do I want to use this expression, and what will make it most useful for this purpose?) and seeing that the basic rules of arithmetic provide a rationale for all legitimate manipulations of polynomial expressions. (NCTM 2009, p. 33)

In this section, we will examine some examples of how "mindful manipulation" can be used in two of the basic polynomial manipulation techniques in high school algebra: factoring and completing the square.

## Factoring

Factoring techniques, especially those that focus on quadratic polynomials, can evolve into a mélange of special-purpose methods that have little utility outside the confines of stylized practice problems. Yet, some factoring methods can be introduced in elementary algebra that extend to polynomials of any degree, preview important ideas in higher algebra, and give students a chance to develop the habits of reasoning about calculations in algebraic structures.

Factoring a quadratic polynomial $x^2 + bx + c$ (with leading coefficient 1) amounts to finding numbers $r$ and $s$ such that

$$r + s = b \quad \text{and}$$
$$rs = c.$$

This "sum-product" approach is important for at least three reasons:
1. Algebra teachers report that students understand the method and can get up to speed with it in one or two class periods.
2. The method can be used later to derive the quadratic formula, and it helps students interpret the *form* of the expressions that result from applying the quadratic formula to the general equation $ax^2 + bx + c = 0$, whose roots are then

$$\frac{-b + \sqrt{b^2 - 4ac}}{2a}, \quad \text{and} \quad \frac{-b - \sqrt{b^2 - 4ac}}{2a}.$$

3. This method previews a much more general correspondence between the coefficients of a polynomial equation and the symmetric functions of its roots, a correspondence that is useful in all parts of algebra and its applications (Barbeau 1989; Cuoco 2005; Graham, Knuth, and Patashnik 1989; Wilf 1994).

However, although most algebra 1 students can master this skill of factoring *monic* (leading coefficient of 1) quadratics by the sum-product approach, many have difficulty when the leading coefficient is not 1. The habit of purposefully transforming expressions can be used to advantage here to hide the complexity posed by a leading coefficient not equal to 1.

One method for reducing the complexity leads to a general-purpose technique that can be applied to polynomials of any degree. As an example, consider

> The difficulty of factoring quadratics that are not monic has led to various methods for "trying all the combinations." Unfortunately, most of these methods have no application outside the factorization of quadratics.

$$4x^2 + 36x + 45.$$

We can think of this expression as a "quadratic in $2x$" if we "chunk" its terms and write it as

$$(2x)^2 + 18(2x) + 45.$$

We can even replace $2x$ by some symbol—say, $z$—and write the quadratic as

$$z^2 + 18z + 45.$$

We can factor this expression by the sum-product method:

$$(z + 15)(z + 3)$$

Replacing $z$ by $2x$ gives us the factorization of the original quadratic.

The coefficients in this example were especially suited to this technique. What if they are not so suited? Suppose that we are faced with factoring something like this:

$$6x^2 + 11x - 10$$

We can reason through the following steps:

1. Multiply the polynomial by 6 to make the leading coefficient a perfect square, remembering that we have to divide by 6 at some point to get back where we started:

$$6(6x^2 + 11x - 10) = 36x^2 + 66x - 60$$

2. The expression $36x^2 + 66x - 60$ is a quadratic in $6x$; let $z = 6x$, so the expression becomes monic:

$$z^2 + 11z - 60$$

3. This expression factors by the sum-product method:

$$(z + 15)(z - 4)$$

4. But $z = 6x$, so we have

$$(6x + 15)(6x - 4).$$

5. Factor out common factors—3 from the first binomial and 2 from the second— producing

$$6(2x + 5)(3x - 2).$$

6. Divide by 6 to obtain the factorization of the original polynomial.

### In the classroom

The following vignette depicts a discussion in an algebra 1 class that has been studying quadratic factoring for a day or two.

*Mrs. Smith:* We've learned that by "chunking" we are able to factor many more quadratic equations than those with a leading coefficient of one. For example, look at this quadratic expression and think about how you might chunk the terms to make it easier to factor [*writes on the board*]:

$$9x^2 + 30x - 24$$

How might you approach factoring this example?

*Todd:* Well, I think we should look at the first term…. I would write $9x^2$ as $(3x)^2$. Then we could write $30x$ as $10(3x)$.

*Mrs. Smith:* Why don't you rewrite $30x$ as $3(10x)$? Isn't that equivalent?

| | |
|---|---|
| *Todd:* | It's equivalent, but I need to match my chunk in the second term with the chunk in the first term so that I can make the problem easier to factor. |
| *Mrs. Smith:* | OK. To help summarize and move us forward, let me do the substitution. Using your chunking idea, we get |

$$(3x)^2 + 10(3x) - 24.$$

Let $z = 3x$, and then we can rewrite the expression as $z^2 + 10z - 24$. Because I know that you all are able to factor that, I'll leave the details for you to finish later. Let's move to another example. Factor this one [*writes again*]:

$$6x^2 + 11x - 10$$

[*Mrs. Smith gives students time to work in pairs.*]

| | |
|---|---|
| *Mrs. Smith:* | Let's come back together and talk about this problem as a class. You all seem to be struggling. Tell me what you are trying to do to factor the expression. |
| *Anna:* | Jim and I have been trying to find a chunk so that we can rewrite the first and second term. We can come up with different ways to rewrite $6x^2$, but we can't seem to find any way to rewrite $11x$. |
| *Mrs. Smith:* | Why might $11x$ be special? |
| *Debbie:* | Well, 11 is a prime number, so the only factors are 11 and 1. |
| *Mrs. Smith:* | Adam, you have a comment? |
| *Adam:* | In each of the problems that we did, we chunked so that the first term was a perfect square. In our last problem, $9x^2$ was a perfect square, but $6x^2$ is not a perfect square. |
| *Mrs. Smith:* | Adam, that's a very interesting observation. Can we make it a perfect square? |
| *Katie:* | Mrs. Smith, if we multiply each term by 6 then the first term would be a perfect square. You would end up with $36x^2 + 66x - 60$. |
| *Amy:* | Wait! You can't just multiply something by 6. |
| *Mrs. Smith:* | Well, let's just remember we multiplied by 6, and we'll deal with it later. |
| *Todd:* | Then we can rewrite the expression as $(6x)^2 + 11(6x) - 60$. We can do a substitution like before by letting $z = 6x$ and get $z^2 + 11z - 60$. |
| *Mrs. Smith:* | Can we factor that? |
| *Frank:* | Yes! You get $(z + 15)(z - 4)$. Then when you substitute back the $6x$, you get $(6x + 15)(6x - 4)$. We did it! |
| *Mrs. Smith:* | I have a concern. We multiplied the expression by 6. Mathematically, that changes the value of the expression. That is not a "legal" mathematical move. Is there any way we can undo what we did to make the factoring easier? |
| *Frieda:* | Mrs. Smith, what if we divide our factors by 6? |
| *Mrs. Smith:* | If we divide each factor by 6, then we are dividing by 36. However, if we can somehow factor out a 6 from the entire expression, that would work for us. |

*Jim:* We can factor out a 3 from $(6x + 15)$, and we can factor out a 2 from $(6x - 4)$. Then the 3 times 2 gets us a factor of 6.

*Mrs. Smith:* Great idea, Jim! Then what we get is [*writes on the board*]

$$(3(2x + 5))(2(3x - 2)), \text{ or } 6(2x + 5)(3x - 2).$$

There is the 6 that we originally multiplied our expression by, so after dividing by 6, our factored quadratic is $(2x + 5)(3x - 2)$.

This method has the benefit of reducing a complicated problem to a problem that is simpler to solve. Letting $z = a_n x$ results in a monic polynomial in $z$ —and saves several days of class time. It is also a general-purpose tool that can be applied to scale a polynomial of any degree so that the transformed polynomial has a leading coefficient of 1. For cubics, the procedure is to scale by the square of the leading coefficient. In general, if

$$f(x) = a_n x^n + a_{n-1} x^{n-1} + \cdots + a_1 x + a_0,$$

then

$$a_n^{n-1} f(x) = (a_n x)^n + a_{n-1}(a_n x)^{n-1} + \cdots + a_n^{n-2} a_1 (a_n x) + a_n^{n-1} a_0.$$

This method is just one tool in an extensive collection of tools that can be used to turn polynomial equations into certain canonical forms that might be easier to solve. One way to think about this method is that it executes a *change of variable*, transforming the quadratic in an invertible way, as in the example above, where the variable is $6x$ instead of $x$.

## Completing the square

The change-of-variable idea can be used to put the topic of completing the square into a larger context. In high school, "completing the square" refers to a method for rewriting a quadratic polynomial so that it has no linear term. In *Focus in High School Mathematics: Reasoning and Sense Making*, example 8, Squaring It Away (pp. 36–37), presents a classroom scene that might unfold when students are asked to solve the equation $x^2 + 10x = 144$. The students explore a geometric method for writing $x^2 + 10x - 144$ as $(x + 5)^2 - 169$. If we let $z = x + 5$ (change the variable to $x + 5$), we can write the original polynomial as $z^2 - 169$, a polynomial in $z$ with no linear term.

How did we know to make the substitution $z = x + 5$? A typical method is to add and subtract half the coefficient of the linear term, but this process frequently seems mysterious to many students. We could have used several methods to determine this substitution; one that involves the mindful manipulation habit is described in more detail in the following:

1. We want to replace $x$ by $z = x - m$ so that $x^2 + 10x - 144$, when expressed as a polynomial in $z$, has no linear term—it looks like $z^2 + p$ for some number $p$. That is, we want

$$x^2 + 10x - 144 = (x - m)^2 + p.$$

2. We can write this expression as

$$x^2 + 10x - 144 = x^2 - 2mx + m^2 + p.$$

3. Hence, we have

$$-2m = 10 \quad \text{and}$$
$$m = -5.$$

> Two polynomials are equal (as polynomials) if and only if they are equal coefficient by coefficient. This idea is another reflection of the formal perspective—thinking of polynomials as objects in their own right.

4. It follows that $p = -169$, and

$$x^2 + 10x - 144 = (x + 5)^2 - 169.$$

This approach to completing the square has some advantages that could be discussed in an advanced second-year algebra or precalculus class:

1. It places the method in a larger context—the "removal of terms" by a linear change of variable. For example, suppose that we want to replace $x$ by $z = x - m$ in the cubic polynomial

$$x^3 - 6x^2 + 5x - 3$$

so that, when expressed in terms of $z$, the resulting cubic has no square term—it looks like $z^3 + pz + q$. The steps are the same:

- We set out to find $m$ so that

$$x^3 - 6x^2 + 5x - 3 = (x - m)^3 + p(x - m) + q$$

for some numbers $p$ and $q$.
- We can write this expression as

$$x^3 - 6x^2 + 5x - 3 = x^3 - 3mx^2 + \text{terms of lower degree}.$$

- It follows that $-3m = -6$, and hence $m = 2$. So $z = x - 2$ is the change of variable that will allow us to remove the square term. A little more work shows us that

$$x^3 - 6x^2 + 5x - 3 = (x - 2)^3 - 7(x - 2) - 9.$$

2. This method can be used to remove the penultimate (next-to-highest degree) term in a polynomial of any degree. In fact, suppose that

$$f(x) = x^n + a_{n-1}x^{n-1} + \cdots + a_1 x + a_0.$$

If $z = \left( x + \dfrac{a_{n-1}}{n} \right)$, then $f(x)$, when expressed as a polynomial in $z$, will have no term of degree $n - 1$. To see this, we proceed as above, letting

$$x^n + a_{n-1}x^{n-1} + \cdots + a_1 x + a_0 = (x - m)^n + 0(x - m)^{n-1} + p_{m-2}(x - m)^{m-2} \ldots$$

We expand the right-hand side by applying the binomial theorem and equate the coefficients of $x^{n-1}$. That leads to a linear equation in $m$,

$$-nm = a_{n-1},$$

> Note that $f(x)$ is monic here. Having a monic polynomial is not a real restriction on the method, because a non-monic polynomial can be transformed into a monic one through another change of variable.

so that $m = -\dfrac{a_{n-1}}{n}$.

The method generalizes further. The same technique can be used to remove the term of degree $n-2$, but this involves solving a quadratic equation for $m$. Removing the term of degree $n-3$ requires the solution of a cubic, and so on. Removing the constant term requires solving an equation of degree $n$—in fact, it requires the solution of $f(m) = 0$.

The habit of judiciously changing a variable—especially completing the square by linear substitution—lies somewhat outside many algebra curricula in use today. However, there are good reasons for including it:

1. It builds a "keen eye" for seeing when a change of variable will simplify a calculation—a useful algebraic habit in many parts of mathematics.

2. It offers the experience of reasoning about calculations with algebraic expressions while helping to develop the habit of "mindful manipulation." The quadratic factoring example is perfectly tractable in algebra 1.

3. A combination of the two methods—transforming polynomials so that they are monic and missing the penultimate term—is a key technique in several parts of algebra, including—

   • the derivation of Cardano's formula for solving cubic equations; and
   • the complete classification of quadratic and cubic polynomials. For example, combining the methods helps in showing that—

      ▫ every quadratic can be transformed to the form $z^2 + p$;
      ▫ every cubic can be transformed to one of three forms: $z^3 + p$, $z^3 + z + p$, or $z^3 - z + p$.

> Cardano's formula is to cubic equations as the quadratic formula is to quadratic equations—an algorithm that expresses the roots of a cubic equation as algebraic expressions of its coefficients. It is of little use in numerical computations, but it has considerable theoretical and historical significance (see Cuoco [2005]).

## Extending the idea

Students in algebra 1 know that there are certain standard ways of writing polynomials. One "normal form" is to write a polynomial as a sum of monomials, arranged from highest degree to lowest. These conventions are built into most CAS environments (see fig. 3.1).

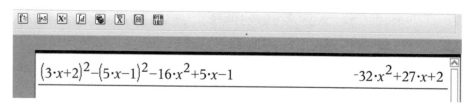

$$(3\cdot x+2)^2-(5\cdot x-1)^2-16\cdot x^2+5\cdot x-1 \qquad -32\cdot x^2+27\cdot x+2$$

Fig. 3.1. A CAS often returns the normal form of a polynomial.

Any polynomial can be put into normal form by using only the basic rules of algebra. Sometimes, though, it is useful to write polynomials in other ways. For example, in the previous discussion of completing the square, we could have formulated the problem of removing the penultimate term in a slightly more general way:

> Some CAS environments report a polynomial as a sum of monomials, listed from lowest to highest degree.

> Given a polynomial $f(x)$, how can we find a number $m$ so that we can write $f(x)$ as powers of $(x - m)$, with no penultimate term?

Expressing polynomials as powers not of $x$, but of some chunk like $x - 2$, can reveal hidden meaning. For example, the fact (derived on page 53) that

$$x^3 - 6x^2 + 5x - 3 = (x - 2)^3 - 7(x - 2) - 9$$

tells us several things:

1. The value of the polynomial when $x = 2$ is $-9$.

2. The remainder when the polynomial is divided by $(x - 2)$ is also $-9$.

3. The remainder when the polynomial is divided by $(x - 2)^2$, or even $(x - 2)^3$, is $-7(x - 2) - 9$.

The remainder theorem from algebra 2 says that statements (1) and (2) will always be related in this way.

We chose to expand in powers of $x - 2$ in this example to remove the quadratic term in the representation, but we could expand in powers of $x - a$ for any number $a$:

$$x^3 - 6x^2 + 5x - 3 = (x - 3)^3 + 3(x - 3)^2 - 4(x - 3) - 15$$

Expressing the polynomial in this way gives all its "local" information—what happens when $x = 3$, what the remainders are when it is divided by powers of $x - 3$, and even how the graphs of $y = x^3 - 6x^2 + 5x - 3$ and $y = (x - 3)^3 + 3(x - 3)^2 - 4(x - 3) - 15$ are related.

Several methods are useful for finding the expansion of a polynomial in powers of $x - a$. This section focuses on one that employs division of polynomials. We have chosen it for several reasons:

1. It showcases the structural similarities between the ordinary integers and the system of polynomials in one variable.

2. It supports the reasoning habit of abstracting regularity from repeated calculations.

3. It previews ideas from calculus and advanced algebra.

The idea of expressing a polynomial written as a combination of powers of $x$ in terms of powers of, say, $x - 3$, is reminiscent of the arithmetic problem of "converting bases." Consider a problem of this type:

> Given the integer 325 expressed as a combination of powers of 10, how can we write this integer as a combination of powers of 7 to get its "base 7" representation?

One method used in many middle school programs proceeds like this:

$$325 = 7 \cdot 46 + 3,$$
$$46 = 7 \cdot 6 + 4,$$
$$6 = 7 \cdot 0 + 6.$$

Putting these all together gives

$$325 = 3 + 7(46)$$
$$= 3 + 7(4 + 7 \cdot 6)$$
$$= 3 + 7 \cdot 4 + 7^2 \cdot 6.$$

> The expansion of 325 as a combination of powers of 7 ($3 + 7 \cdot 4 + 7^2 \cdot 6$) exactly reflects the meaning of the digits in the decimal expansion of a number, where the digits are successive remainders when dividing by 10. It is customary to write this expansion as $(643)_7$, following the "highest powers to the left" convention.

This is the base 7 expansion for 325; it shows how 325 is represented as a combination of powers of 7.

The situation is exactly the same for polynomials. Essentially, we are converting from one "base" (powers of $x$) to another (powers of $x - a$). Using this method as inspiration, let's suppose that we want to write $f(x) = x^4 - 5x^3 + 3x - 1$ as a polynomial in $x - 3$. We can execute the same process that we did for the base 7 expansion, this time with polynomials.

First, we divide $f(x)$ by $x - 3$; see figure 3.2. The result of the division says that

$$f(x) = -46 + (x-3)\underbrace{\left(x^2 - 2x^2 - 6x - 15\right)}_{q(x)},$$

where $q(x)$ is the quotient and –46 is the remainder.

> When it is possible to "execute the same process in two different systems," there is a good chance that the systems have some underlying structural similarity.

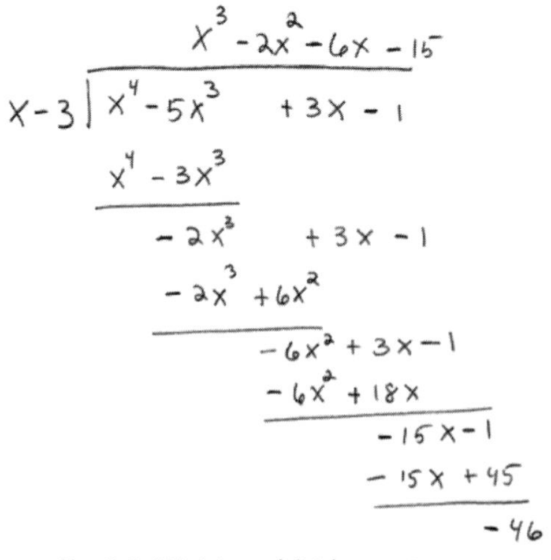

$$
\begin{array}{r}
x^3 - 2x^2 - 6x - 15 \\
x-3 \overline{\smash{\big)}\ x^4 - 5x^3 \qquad + 3x - 1} \\
\underline{x^4 - 3x^3} \\
-2x^3 \qquad + 3x - 1 \\
\underline{-2x^3 + 6x^2} \\
-6x^2 + 3x - 1 \\
\underline{-6x^2 + 18x} \\
-15x - 1 \\
\underline{-15x + 45} \\
-46
\end{array}
$$

Fig. 3.2. Division of $f(x)$ by $x - 3$

> Let $x = 3$ to find that $f(3) = -46$.

Next, we divide $q(x) = x^3 - 2x^2 - 6x - 15$ by $x - 3$; see figure 3.3. This division says that

$$q(x) = -24 + \underbrace{(x-3)(x^2 + x - 3)}_{q_1(x)}.$$

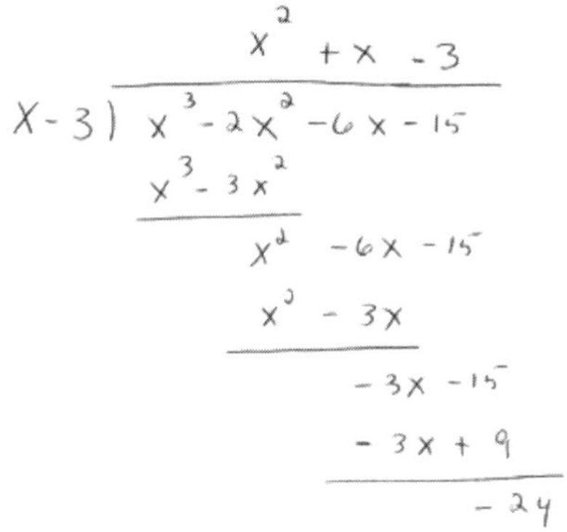

$$
\begin{array}{r}
x^2 + x - 3 \\
x - 3 \overline{)\ x^3 - 2x^2 - 6x - 15} \\
\underline{x^3 - 3x^2} \\
x^2 - 6x - 15 \\
\underline{x^2 - 3x} \\
-3x - 15 \\
\underline{-3x + 9} \\
-24
\end{array}
$$

Fig. 3.3. Division of $q(x)$ by $x - 3$

Now we divide $q_1(x) = x^2 + x - 3$ by $x - 3$; see figure 3.4. This division says that

$$q_1(x) = 9 + \underbrace{(x-3)(x + 4)}_{q_2(x)}.$$

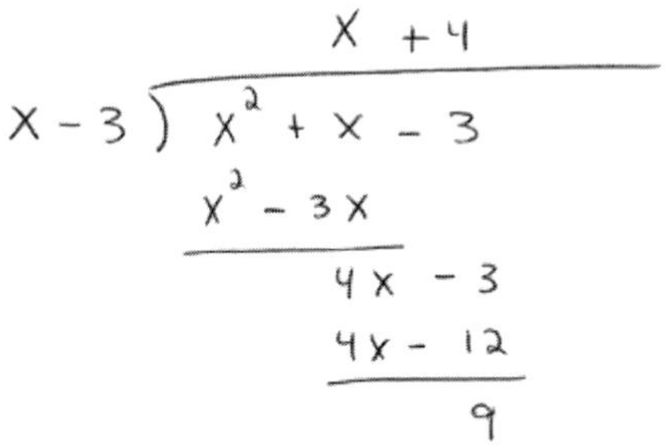

$$
\begin{array}{r}
x + 4 \\
x - 3 \overline{)\ x^2 + x - 3} \\
\underline{x^2 - 3x} \\
4x - 3 \\
\underline{4x - 12} \\
9
\end{array}
$$

Fig. 3.4. Division of $q_1(x)$ by $x - 3$

Next, we divide $q_2(x) = x + 4$ by $x - 3$; see figure 3.5. This division says that

$$q_2(x) = 7 + \underbrace{(x-3)(1)}_{q_3(x)}.$$

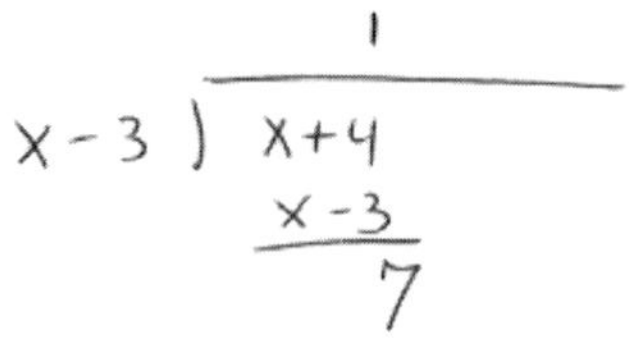

$$
\begin{array}{r}
1\phantom{xxxx} \\
x-3 \overline{)\ x+4\phantom{xx}} \\
\underline{x-3\phantom{x}} \\
7
\end{array}
$$

Fig. 3.5. Division of $q_2(x)$ by $x-3$

Now, we can put it all together:

$$
\begin{aligned}
f(x) &= -46 + (x-3)\underbrace{\left(x^3\ \ 2x^2 - 6x - 15\right)}_{q(x)} \\
&= -46 + (x-3)(-24 + (x-3)(x^2 + x - 3)) \\
&= -46 - 24(x-3) + (x-3)^2\left(\underbrace{x^2 + x - 3}_{q_1(x)}\right) \\
&= -46 - 24(x-3) + (x-3)^2\left(9 + (x-3)(x+4)\right) \\
&= -46 - 24(x-3) + 9(x-3)^2 + (x-3)^3\left(\underbrace{x+4}_{q_2(x)}\right) \\
&= -46 - 24(x-3) + 9(x-3)^2 + (x-3)^3\left(7 + (x-3)\right) \\
&= -46 - 24(x-3) + 9(x-3)^2 + 7(x-3)^3 + (x-3)^4
\end{aligned}
$$

In older algebra books, such calculations were often carried out with a pretty variant of synthetic division, as shown in figure 3.6.

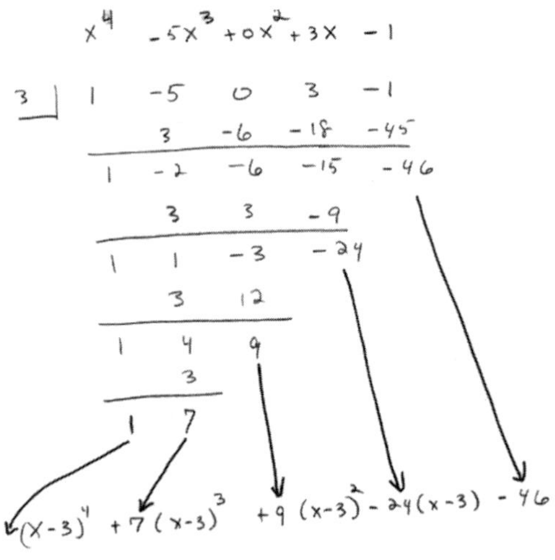

Fig. 3.6. Repeated synthetic division of $f(x)$ by $x-3$

A CAS can be used in several ways to reduce the computational overhead in these calculations. The following are examples:

- Most systems have the ability to calculate the remainder when dividing one polynomial by another. The remainders and partial quotients in the calculation on page 58 can be obtained with these CAS functions; see figure 3.7a.

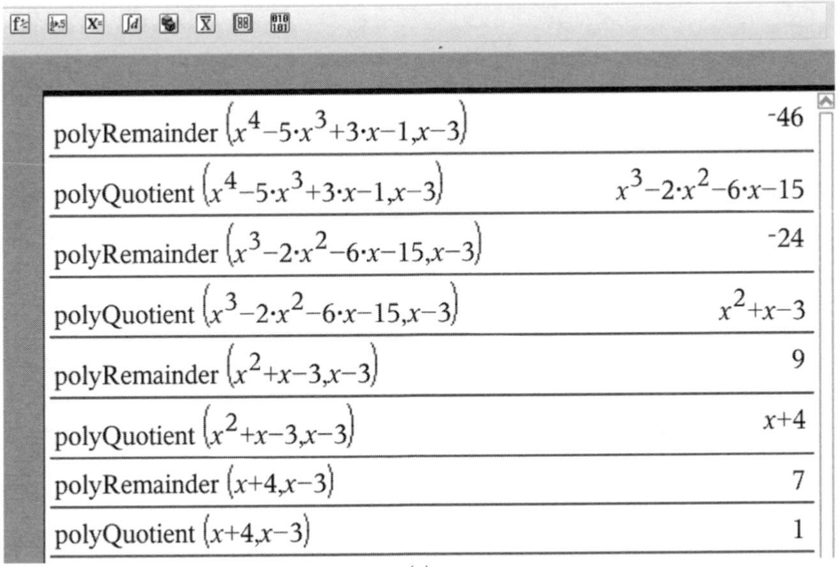

| | |
|---|---|
| polyRemainder $\left(x^4-5{\cdot}x^3+3{\cdot}x-1, x-3\right)$ | -46 |
| polyQuotient $\left(x^4-5{\cdot}x^3+3{\cdot}x-1, x-3\right)$ | $x^3-2{\cdot}x^2-6{\cdot}x-15$ |
| polyRemainder $\left(x^3-2{\cdot}x^2-6{\cdot}x-15, x-3\right)$ | -24 |
| polyQuotient $\left(x^3-2{\cdot}x^2-6{\cdot}x-15, x-3\right)$ | $x^2+x-3$ |
| polyRemainder $\left(x^2+x-3, x-3\right)$ | 9 |
| polyQuotient $\left(x^2+x-3, x-3\right)$ | $x+4$ |
| polyRemainder $\left(x+4, x-3\right)$ | 7 |
| polyQuotient $\left(x+4, x-3\right)$ | 1 |

(a)

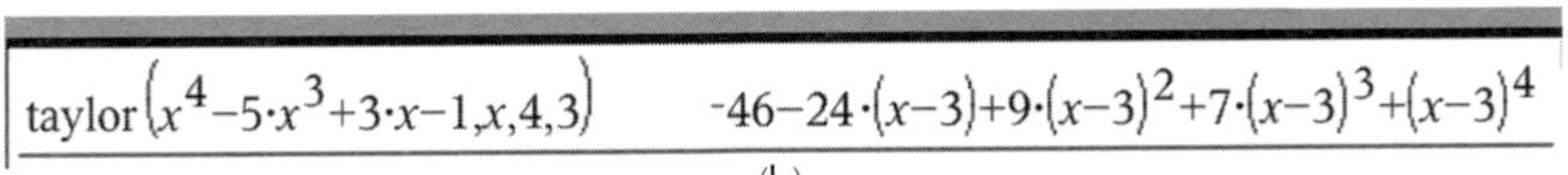

| | |
|---|---|
| taylor $\left(x^4-5{\cdot}x^3+3{\cdot}x-1, x, 4, 3\right)$ | $-46-24{\cdot}(x-3)+9{\cdot}(x-3)^2+7{\cdot}(x-3)^3+(x-3)^4$ |

(b)

Fig. 3.7. CAS screenshots of (a) $f(x)$ divided by $x - 3$ and (b) the Taylor expansion for $f(x)$ about $x = 3$

- The resulting expression of $f(x)$ as a combination of powers of $x - 3$ is the *Taylor expansion* for $f(x)$ about $x = 3$; see figure 3.7b. Taylor expansions are usually derived in calculus, but we have derived one here by purely algebraic means. In the figure, the syntax in the TI-Nspire software is read, "Give me the expansion of $x^4 - 5x^3 + 3x -1$; the variable is $x$, the degree is 4, and the expansion is in terms of $x - 3$."

$$\text{taylor} \quad \left(x^4 - 5x^3 + 3x - 1, \quad x, \quad 4, \quad 3\right)$$
$$\downarrow \qquad\qquad \downarrow \quad\quad \downarrow \qquad\quad \downarrow$$
$$\text{polynomial} \quad \text{variable} \quad \text{degree} \quad \text{in terms of } (x - 3)$$

> The calculus derivation of the expansion uses the function perspective. The long-division approach treats the polynomial as a formal algebraic object.

# Algebra as Bookkeeping

In the preceding section, we focused on transforming polynomials in various ways so that we could gain information about their structure and behavior. In this section, we examine another application of the formal perspective: using polynomial calculations as counting devices. Consider an example involving distributions of the sums in tosses of number cubes.

Many elementary and middle school curricula ask students to experiment with probability. In one of the most popular experiments, students toss number cubes while exploring the following kinds of questions:

1.  When tossing a single cube, how many different numbers can show up? What is the most likely number?

2.  When tossing two number cubes, how many different pairs can show up? How many different sums can show up? What is the most likely sum?

The possible sums for two number cubes (problem 2 above) are 2, 3, …, 12. A common representation shows what can happen with each cube, listing all 36 possibilities for their sum; see the table in figure 3.8.

| Cube 1 →<br>Cube 2 ↓ | 1 | 2 | 3 | 4 | 5 | 6 |
|---|---|---|---|---|---|---|
| 1 | 2 | 3 | 4 | 5 | 6 | 7 |
| 2 | 3 | 4 | 5 | 6 | 7 | 8 |
| 3 | 4 | 5 | 6 | 7 | 8 | 9 |
| 4 | 5 | 6 | 7 | 8 | 9 | 10 |
| 5 | 6 | 7 | 8 | 9 | 10 | 11 |
| 6 | 7 | 8 | 9 | 10 | 11 | 12 |

Fig. 3.8. A table of possibilities for the sums when two number cubes are rolled

The table shows not only that 7 is the most likely sum, but also how the other sums are *distributed*. This distribution can be read from the table by collecting the like occurrences of each integer, as illustrated in figure 3.9.

| Cube 1 →<br>Cube 2 ↓ | 1 | 2 | 3 | 4 | 5 | 6 |
|---|---|---|---|---|---|---|
| 1 | 2 | 3 | 4 | 5 | 6 | 7 |
| 2 | 3 | 4 | 5 | 6 | 7 | 8 |
| 3 | 4 | 5 | 6 | 7 | 8 | 9 |
| 4 | 5 | 6 | 7 | 8 | 9 | 10 |
| 5 | 6 | 7 | 8 | 9 | 10 | 11 |
| 6 | 7 | 8 | 9 | 10 | 11 | 12 |

| Sum | 2 | 3 | 4 | 5 | 6 | 7 | 8 | 9 | 10 | 11 | 12 |
|---|---|---|---|---|---|---|---|---|---|---|---|
| Number of occurrences | 1 | 2 | 3 | 4 | 5 | 6 | 5 | 4 | 3 | 2 | 1 |

Fig. 3.9. Collecting like occurrences in a table of possible sums and making a frequency table of the results

What does this work have to do with algebra? Well, students often multiply two polynomials by using a similar tabular method for keeping track of all the pairs of terms. For example, to multiply $x^2 - 3x + 2$ by $2x^2 + 5x - 1$, they can arrange the calculation in a table, as in figure 3.10.

| Poly 1 → Poly 2 ↓ | $x^2$ | $-3x$ | 2 |
|---|---|---|---|
| $2x^2$ | $2x^4$ | $-6x^3$ | $4x^2$ |
| $5x$ | $5x^3$ | $-15x^2$ | $10x$ |
| $-1$ | $-x^2$ | $3x$ | $-2$ |

Fig. 3.10. A table representing the terms in the polynomial multiplication $(x^2 - 3x + 2)(2x^2 + 5x - 1)$

The students can read the product from the table by collecting the like terms, as in figure 3.11. This process yields the result

$$(x^2 - 3x + 2)(2x^2 + 5x - 1) = 2x^4 - x^3 - 12x^2 + 13x - 2.$$

| Poly 1 → Poly 2 ↓ | $x^2$ | $-3x$ | 2 |
|---|---|---|---|
| $2x^2$ | $2x^4$ | $-6x^3$ | $4x^2$ |
| $5x$ | $5x^3$ | $-15x^2$ | $10x$ |
| $-1$ | $-x^2$ | $3x$ | $-2$ |

| Term | $x^4$ | $x^3$ | $x^2$ | $x$ | 1 |
|---|---|---|---|---|---|
| Number of occurrences | 2 | $-1$ | $-12$ | 13 | $-2$ |

Fig. 3.11. Collecting like terms yields the product of polynomials.

> This method of multiplying polynomials works exactly the same as the "lattice method" for multiplying base ten numbers (but in that situation, it must be supplemented with regrouping).

You can explore an interesting problem with your class:

What polynomials could be multiplied to model the distribution of sums when two number cubes are thrown?

The students might experiment, trying to find such polynomials by using the table in figure 3.8. One calculation that mirrors the distribution table exactly is

$$(x + x^2 + x^3 + x^4 + x^5 + x^6)^2,$$

as shown in figure 3.12.

| Poly 1→ Poly 2↓ | $x$ | $x^2$ | $x^3$ | $x^4$ | $x^5$ | $x^6$ |
|---|---|---|---|---|---|---|
| $x$ | $x^2$ | $x^3$ | $x^4$ | $x^5$ | $x^6$ | $x^7$ |
| $x^2$ | $x^3$ | $x^4$ | $x^5$ | $x^6$ | $x^7$ | $x^8$ |
| $x^3$ | $x^4$ | $x^5$ | $x^6$ | $x^7$ | $x^8$ | $x^9$ |
| $x^4$ | $x^5$ | $x^6$ | $x^7$ | $x^8$ | $x^9$ | $x^{10}$ |
| $x^5$ | $x^6$ | $x^7$ | $x^8$ | $x^9$ | $x^{10}$ | $x^{11}$ |
| $x^6$ | $x^7$ | $x^8$ | $x^9$ | $x^{10}$ | $x^{11}$ | $x^{12}$ |

Fig. 3.12. An expansion table for $(x + x^2 + x^3 + x^4 + x^5 + x^6)^2$

One way to encourage students to develop this insight is to ask a series of questions such as the following:

1. How would you determine the coefficient of $x^7$ in the expansion of $(x + x^2 + x^3 + x^4 + x^5 + x^6)^2$?
2. How would you determine the coefficient of $x^8$ in the product?
3. How would you determine the coefficient of $x^9$ in the product?
4. How do you determine the coefficient of any term $x^n$ in the product?

One frequent answer to the last question is, "The coefficient of $x^n$ is the number of ordered pairs of integers, each between 1 and 6, that add up to $n$." This is exactly the number of ways of getting a sum of $n$ in tossing two number cubes.

What about the coefficient of $x^n$ in the expansion of $(x + x^2 + x^3 + x^4 + x^5 + x^6)^3$? A careful analysis of the calculations leads to a similar result: The coefficient of $x^n$ in the expansion of $(x + x^2 + x^3 + x^4 + x^5 + x^6)^3$ is the number of ordered triples of integers, each between 1 and 6, that add up to $n$. This is exactly the number of ways of getting a sum of $n$ in tossing three number cubes.

Indeed, we get a general result:

**Theorem:** The coefficient of $x^n$ in the expansion of $(x + x^2 + x^3 + x^4 + x^5 + x^6)^k$ is the number of ways of getting a sum of $n$ in tossing $k$ number cubes.

A CAS can give students the distribution data, as illustrated in figure 3.13.

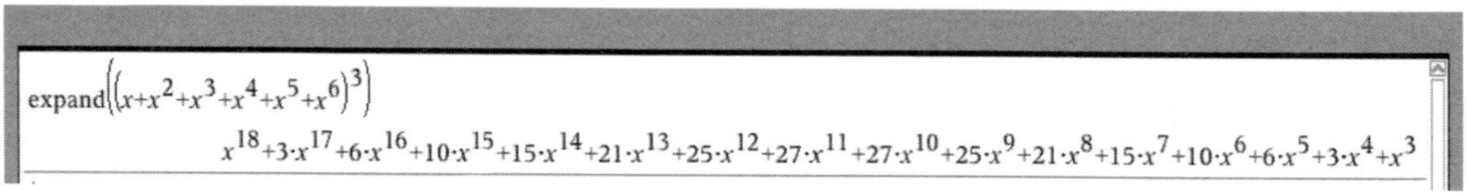

expand$\left((x+x^2+x^3+x^4+x^5+x^6)^3\right)$

$x^{18}+3 \cdot x^{17}+6 \cdot x^{16}+10 \cdot x^{15}+15 \cdot x^{14}+21 \cdot x^{13}+25 \cdot x^{12}+27 \cdot x^{11}+27 \cdot x^{10}+25 \cdot x^9+21 \cdot x^8+15 \cdot x^7+10 \cdot x^6+6 \cdot x^5+3 \cdot x^4+x^3$

Fig. 3.13. A CAS screenshot showing the expansion of $(x + x^2 + x^3 + x^4 + x^5 + x^6)^3$

One interesting feature of the distribution is that the first few coefficients might give students the impression that the sequence will follow triangular numbers. But that pattern breaks down at 12, raising two interesting questions:

1.  Why do triangular numbers show up in the first place?
2.  Why does the pattern break down?

An analysis of both the algebra and the context can provide answers.

Students now have a laboratory in which they can experiment, looking at properties of the distributions and finding how they are mirrored in the algebra. As with the area formulas in chapter 1, algebra informs and is informed by the situations that it models.

Possible questions for your students to investigate include the following:

1.  Thirty-six "outcomes" are possible when you throw two number cubes. How could you use the algebraic model for the distribution of sums to get this number?

> What substitution can you make in a polynomial to get the sum of its coefficients?

2.  If you toss two number cubes, can you get more even sums than odd sums? More odd sums than even? Are the numbers of odd and even sums that you can get the same?

    a.  How could you prove that your answer is correct by using number cubes?
    b.  How could you prove that your answer is correct by using polynomials?

3.  Use the algebra of "distribution polynomials" to obtain—

    a.  the number of possible "outcomes" that you can get when you toss three number cubes;
    b.  the difference between the number of sums that are even and the number of sums that are odd that you can get when you toss three number cubes;
    c.  the most likely sum or sums that you can get when you toss three number cubes.

4.  What is the distribution of the sums that you can get when you toss four number cubes?

5.  Use the algebra of "distribution polynomials" to obtain—

    a.  the number of possible "outcomes" that you can get when you toss four number cubes;
    b.  the difference between the number of sums that are even and the number of sums that are odd that you can get when you toss four number cubes;
    c.  the sum or sums that you are most likely to get when you toss four number cubes.

6.  Find a formula in terms of $n$ for the sum that someone is most likely to get when tossing $n$ number cubes.

You might pose more challenging questions:

7.  Suppose you have two number cubes. One cube is "fair," and the other has faces labeled $\{1, 1, 3, 4, 5, 6\}$. What polynomial calculation gives you the distribution of sums that you can get in tossing the number cubes? What is the new distribution?
8.  Suppose you have three number cubes. Two number cubes are fair, and the other has faces

labeled $\{2, 2, 4, 4, 5, 8\}$. What is the new distribution of sums that you can get in tossing these number cubes? What sums can you not get with these cubes that you can get with three fair number cubes? What new sums can show up with the new cubes?

9. What is the distribution of sums that you can get when you toss three number cubes labeled

$$\{0, 2, 3, 4, 5, 5\}, \quad \{0, 1, 1, 2, 2, 2\}, \quad \text{and} \quad \{1, 2, 3, 6, 6, 6\} \ ?$$

Exploring these types of questions provides students with an opportunity to develop their skill in using algebraic calculations to mirror various situations that involve counting and enumeration. This work is important for a number of reasons, including the following:

1. It helps students develop expertise at both picking algebraic calculations apart and designing algebraic calculations to model situations.
2. It strengthens in students the habit of purposeful transformation.
3. It exposes students to methods that are the basis for a whole field of applied algebra—the theory of *generating functions*—that finds applications in all parts of mathematics and science (Wilf 1994).

We have chosen to focus on the number cubes example because of the easy entry point that it offers students. Other contexts in high school mathematics lend themselves to the generating function technique. Examples of problems in two such contexts follow:

1. How many ways you can make change for a dollar by using only the standard coins?
2. The post office has only 5- and 8-cent stamps today. How many amounts less than 100 can you make in at least one way? Can you make any amounts in two ways? More than two ways?

## Extending the idea

The ability to use polynomial calculations to model the distribution of sums in tosses of number cubes gives students a new tool—formal algebra—to use to answer questions about the distributions. The examples that accompany some of the ideas in this section might be interesting for students in more advanced classes, such as precalculus.

For instance, if we want an explicit formula for the number of ways that we can get sums of $n$ when tossing $k$ number cubes, we could look for an explicit formula for the coefficient of $x^n$ in

$$(x + x^2 + x^3 + x^4 + x^5 + x^6)^k.$$

One route for investigation is to look closely at the structure of the polynomial

$$x + x^2 + x^3 + x^4 + x^5 + x^6.$$

Factoring out an $x$, we have

$$x + x^2 + x^3 + x^4 + x^5 + x^6 = x(1 + x + x^2 + x^3 + x^4 + x^5).$$

The second factor shows up in example 6, Distribute Thoroughly (pp. 33–34), in *Focus in High School Mathematics: Reasoning and Sense Making,* and plays an important role in many high school contexts, including geometric series and De Moivre's theorem. Two useful algebraic identities in these contexts are given below:

1.
$$(1 - x)(1 + x + x^2 + x^3 + \cdots + x^{n-1}) = 1 - x^n,$$

   so that

$$1 + x + x^2 + x^3 + \cdots + x^{n-1} = \frac{1 - x^n}{1 - x}.$$

2.
$$\frac{1}{1-x} = 1 + x + x^2 + x^3 + \ldots$$

> Both (1) and (2) are formal algebraic identities, established by formal algebra. They are therefore true under any substitution for which the expressions make sense—in the first one, $x \neq 1$, and in the second one, $|x| < 1$.

These two facts imply that

$$x + x^2 + x^3 + x^4 + x^5 + x^6 = x\left(\frac{1 - x^6}{1 - x}\right)$$
$$= x\left(1 - x^6\right)\left(1 - x\right)^{-1}$$
$$= x\left(1 - x^6\right)\left(1 + x + x^2 + \ldots\right),$$

so that

$$(x + x^2 + x^3 + x^4 + x^5 + x^6)^k = x^k(1 - x^6)^k(1 + x + x^2 + \ldots)^k.$$

This last expression lends itself to a search for an explicit formula for the coefficient of $x^n$.

Another problem that intrigues students and teachers alike explores the possibility of finding unconventionally labeled number cubes that act as "fair" cubes in tosses. This problem can be stated as follows:

> Is it possible to label the faces of two number cubes with positive integers in a way that is different from putting 1–6 on each—so that the distribution of sums that you can get in tossing the cubes is the same as the distribution that you can get when the number cubes are labeled in the usual way?

The two number cubes do not have to have the same labels, and an integer greater than 6 can appear on a face.

Labeling two 6-sided number cubes with new numbers so that the distribution of sums is the same as that for cubes with the "normal" labeling amounts to rearranging the factors of

$$x^2 \left( \frac{x^6 - 1}{x - 1} \right)^2$$

into two unequal polynomials that each stand for a valid 6-sided cube. So, students would be led to an investigation of the irreducible factors of the polynomial above. Figure 3.14 shows the factorization.

$$\text{factor}\left(\left(x + x^2 + x^3 + x^4 + x^5 + x^6\right)^2\right)$$

$$x^2 \cdot (x+1)^2 \cdot \left(x^2 + x + 1\right)^2 \cdot \left(x^2 - x + 1\right)^2$$

Fig. 3.14. Expressing $(x + x^2 + x^3 + x^4 + x^5 + x^6)^2$ as a product of irreducible factors

For an $m$-sided die, we would need to rearrange the factors of

$$x^2 \left( \frac{x^m - 1}{x - 1} \right)^2 .$$

This leads to an interesting question: How many factors does $x^m - 1$ have over the integers? In the next section, we explore this question.

## Experimenting with Expressions

A computer algebra system is a wonderful tool for helping students begin to see algebraic expressions as "real things." In this section, we look at an activity that aims to assist students in bringing the objects of algebra into their real world. In algebra 1, students learn to factor $x^2 - 1$ and perhaps $x^3 - 1$ and $x^4 - 1$. A list of the factorizations from a CAS (see fig. 3.15) provides data for some very interesting algebraic investigations.

| Define $f(n) = \text{factor}\left(x^n - 1\right)$ | Done |
|---|---|
| $f(2)$ | $(x-1) \cdot (x+1)$ |
| $f(3)$ | $(x-1) \cdot \left(x^2 + x + 1\right)$ |
| $f(4)$ | $(x-1) \cdot (x+1) \cdot \left(x^2 + 1\right)$ |
| $f(5)$ | $(x-1) \cdot \left(x^4 + x^3 + x^2 + x + 1\right)$ |
| $f(6)$ | $(x-1) \cdot (x+1) \cdot \left(x^2 + x + 1\right) \cdot \left(x^2 - x + 1\right)$ |
| $f(7)$ | $(x-1) \cdot \left(x^6 + x^5 + x^4 + x^3 + x^2 + x + 1\right)$ |
| $f(8)$ | $(x-1) \cdot (x+1) \cdot \left(x^2 + 1\right) \cdot \left(x^4 + 1\right)$ |
| $f(9)$ | $(x-1) \cdot \left(x^2 + x + 1\right) \cdot \left(x^6 + x^3 + 1\right)$ |

Fig. 3.15. Factorizations of $x^n - 1$ for $2 \le n \le 9$

Even before investigating the nature of the factors, students can organize the data from the CAS to try to find regularity in the *number* of factors, as in the table below:

| $n$ | number of factors of $x^n - 1$ |
|---|---|
| 1 | 1 |
| 2 | 2 |
| 3 | 2 |
| 4 | 3 |
| 5 | 2 |
| 6 | 4 |
| 7 | 2 |
| 8 | 4 |
| 9 | 3 |

In this situation, students are using the CAS simply to gather data. Along the way, you can ask them to reason about how the number of factors of $x^n - 1$ is related to the number of factors for $x^m - 1$ for $m < n$, by using chunking. For example, $x^8 - 1 = (x^2)^4 - 1$.

If you choose to do this activity with your students, they might come up with a variety of conjectures. Conjectures that we have seen in work with students and teachers include the following:

- There are always at least two factors:

$$x^n - 1 = (x - 1)(x^{n-1} + x^{n-2} + \cdots + x^2 + x + 1)$$

- If $n$ is odd, then there are exactly two factors (but look at $n = 9$).
- OK..., if $n$ is prime, then there are exactly two factors.
- If $n$ is the square of a prime, then there are three factors ($x^9 - 1$, for example).
- If $n$ is the product of two different primes, then there are four factors ($x^{15} - 1$, for example).

Eventually, a general conjecture emerges:

> **Conjecture:** The number of irreducible factors of $x^n - 1$ over the integers is the number of factors of $n$ in positive integers.

This is a rather deep result in algebra, and its proof is most likely beyond most high school students. Still, even without complete proofs, students can benefit from the activity, for several reasons:

1. Within 15 minutes, many students in an algebra 2 class begin to act as if these polynomials and their factors are real things for them.
2. The data can provide students with a nontrivial example of pattern spotting.
3. This context can give rise to questions and investigations that bring students to the frontiers of what is known in mathematics.

## Extending the idea

Taking the investigation to the next level of detail involves looking closely at the degrees of the irreducible factors. The chart in figure 3.16 organizes the data.

| $n$ | Degrees of irreducible factors of $x^n - 1$ | $n$ | Degrees of irreducible factors of $x^n - 1$ |
|---|---|---|---|
| 1 | 1 | 11 | 1, 10 |
| 2 | 1, 1 | 12 | 1, 1, 2, 2, 2, 4 |
| 3 | 1, 2 | 13 | 1, 12 |
| 4 | 1, 1, 2 | 14 | 1, 1, 6, 6 |
| 5 | 1, 4 | 15 | 1, 2, 4, 8 |
| 6 | 1, 1, 2, 2 | 16 | 1, 1, 2, 4, 8 |
| 7 | 1, 6 | 17 | 1, 16 |
| 8 | 1, 1, 2, 4 | 18 | 1, 1, 2, 2, 6, 6 |
| 9 | 1, 2, 6 | 19 | 1, 18 |
| 10 | 1, 1, 4, 4 | 20 | 1, 1, 2, 4, 4, 8 |

Fig. 3.16. Degrees of irreducible factors of $x^n - 1$ for $2 \le n \le 20$

Once again, these data are a treasure trove for making conjectures and finding partial verifications. It turns out that for every factor $d$ of $n$, $x^n - 1$ has one irreducible factor, and the degree of that factor is the number of positive integers less than $d$ that have no common factor with $d$. Looking even deeper into the data on the factors of $x^n - 1$ suggests that all the nonzero coefficients of the irreducible factors are either 1 or –1. However, a counterexample occurs in $x^{105} - 1$, where there is a coefficient of –2.

## Conclusion

These examples lie at the end of the spectrum outlined in *Focus in High School Mathematics: Reasoning and Sense Making*. In this book, we have moved from considering algebraic expressions within the context of geometry to a perspective in which the focus is on the properties of the *operations* and in which the *expressions* are part of a system that is its own context, open to experiment in exactly the same way that dissections and area formulas are open to investigation. Considering algebraic expressions from the perspective of the properties of the operations and as part of a system that is its own context reinforces the idea that polynomials are real objects in a complex system, with all the textured intricacies of physical phenomena. Students who work with algebra in this context get quite close to results of current mathematical research.

In fact, it is known that the coefficients of the irreducible factors of $x^n - 1$ can be made as large as desired by making the exponent $n$ big enough. Indeed, in 1987, Jiro Suzuki proved the following result:

> If $k$ is odd, and if $p_1 < p_2 < \cdots < p_k$ is a "front-loaded" sequence of primes—that is, the sum of the first two in the sequence is greater than the last—and if $n$ is the product of all the primes in the sequence, then one of the irreducible factors of $x^n - 1$ has $-k + 1$ and $-k + 2$ as coefficients.

Suzuki also proved that there is a front-loaded sequence of primes of length $k$ for every odd $k$. Note that $105 = 3 \cdot 5 \cdot 7$. Also, large coefficients show up much earlier than the theorem guarantees.

# Epilogue

The chapters in this book have provided a range of examples that you can implement to offer opportunities for students to engage in reasoning and sense making in the context of high school algebra. We began by considering algebraic expressions that were closely connected to geometric situations, and we moved to considering algebraic expressions that focus on the properties of the *operations* and are part of a system that is its own context, open to experiment in exactly the same way that dissections and area formulas are.

As we mentioned earlier, the examples that we have presented do not represent an exhaustive list of topics to be covered in any specific course or curriculum. In particular, this book is not an algebra curriculum. Rather, it is a set of key algebraic ideas that highlight reasoning and sense making across grades 9–12. The concepts illustrated by the examples in each chapter do not, in and of themselves, guarantee that students will engage in reasoning and sense making in the classroom. The teacher plays a critical role in creating a classroom environment that emphasizes reasoning and sense making as an everyday occurrence.

We hope that you will find the classroom vignettes and examples of student reasoning contained in this book helpful as you reflect on reasoning and sense making in your own classroom, school, or district. A tremendous amount of research has been done over the last twenty years on students' fundamental understanding of central algebraic concepts such as variable, equation, and function. Important resources in this area include *Second Handbook of Research on Mathematics Teaching and Learning* (Lester 2007) and *A Research Companion to "Principles and Standards for School Mathematics"* (Kilpatrick, Martin, and Schifter 2003). Much has also been written about the role of and approaches to algebra in the precollege curriculum (Bednarz, Kieran, and Lee 1996; Coxford 1985; Driscoll 1999; Greenes 2008; Kieran 2004). Whatever approach these authors take, they believe that structuring experiences that emphasize reasoning and sense making is essential to a student's ability to move from the concrete to the abstract and to his or her success in future mathematics courses.

The examples in this book have provided additional detail about the fundamental habits of mind and key elements of algebraic reasoning outlined in *Focus in High School Mathematics: Reasoning and Sense Making* (NCTM 2009). They have demonstrated some of the possibilities for using properties of algebra to illustrate what is happening geometrically, building equations from calculations by "expressing the rhythm" in the calculation, and considering algebraic expressions as objects in their own right. As you reflect on the school mathematics curriculum in the context of reasoning and sense making in algebra, consider the following questions:

- How can the curriculum integrate technology to help students experience reasoning and sense making?

- What are possible ways to modify textbook problems or exercises to elicit and support students' reasoning?

- What kinds of algebraic tasks can stimulate students' reasoning and sense making?

- What are possible ways to structure a classroom environment to maximize the opportunities for evaluating and sharing students' ideas?

- What are possible ways to model the types of thinking that students need to do?

- What types of questions can foster the development of algebraic reasoning in students?

We hope that you have enjoyed the journey—and that you are looking forward to the journey ahead in reasoning and sense making in algebra!

# Appendix

NCTM Standards and Expectations for Grades 9–12

# Algebra
# Standard

| *Instructional programs from prekindergarten through grade 12 should enable all students to—* | **Grades 9–12**<br>**Expectations**<br>**In grades 9–12 all students should—** |
|---|---|
| Understand patterns, relations, and functions | • generalize patterns using explicitly defined and recursively defined functions;<br>• understand relations and functions and select, convert flexibly among, and use various representations for them;<br>• analyze functions of one variable by investigating rates of change, intercepts, zeros, asymptotes, and local and global behavior;<br>• understand and perform transformations such as arithmetically combining, composing, and inverting commonly used functions, using technology to perform such operations on more complicated symbolic expressions;<br>• understand and compare the properties of classes of functitons, including exponential, polynomial, rational, logarithmic, and periodic functions;<br>• interpret representations of functions of two variables. |
| Represent and analyze mathematical situations and structures using algebraic symbols | • understand the meaning of equivalent forms of expressions, equations, inequalities, and relations;<br>• write equivalent forms of equations, inequalities, and systems of equations and solve them with fluency—mentally or with paper and pencil in simple cases and using technology in all cases;<br>• use symbolic algebra to represent and explain mathematical relationships;<br>• use a variety of symbolic representations, including recursive and parametric equations, for functions and relations;<br>• judge the meaning, utility, and reasonableness of the results of symbol manipulations, including those carried out by technology. |
| Use mathematical models to represent and understand quantitative relationships | • identify essential quantitative relationships in a situation and determine the class or classes of functions that might model the relationships;<br>• use symbolic expressions, including iterative and recursive forms, to represent relationships arising from various contexts;<br>• draw reasonable conclusions about a situation being modeled. |
| Analyze change in various contexts | • approximate and interpret rates of change from graphical and numerical data. |

# Number and Operations Standard

| Instructional programs from prekindergarten through grade 12 should enable all students to— | Grades 9–12 Expectations<br>In grades 9–12 all students should— |
|---|---|
| Understand numbers, ways of representing numbers, relationships among numbers, and number systems | • develop a deeper understanding of very large and very small numbers and of various representations of them;<br>• compare and contrast the properties of numbers and number systems, including the rational and real numbers, and understand complex numbers as solutions to quadratic equations that do not have real solutions;<br>• understand vectors and matrices as systems that have some of the properties of the real-number system;<br>• use number-theory arguments to justify relationships involving whole numbers. |
| Understand meanings of operations and how they relate to one another | • judge the effects of such operations as multiplication, division, and computing powers and roots on the magnitudes of quantities;<br>• develop an understanding of properties of, and representations for, the addition and multiplication of vectors and matrices;<br>• develop an understanding of permutations and combinations as counting techniques. |
| Compute fluently and make reasonable estimates | • develop fluency in operations with real numbers, vectors, and matrices, using mental computation or paper-and-pencil calculations for simple cases and technology for more complicated cases;<br>• judge the reasonableness of numerical computations and their results. |

| Measurement Standard | |
| --- | --- |
| *Instructional programs from prekindergarten through grade 12 should enable all students to—* | **Grades 9–12 Expectations**<br>**In grades 9–12 all students should—** |
| Understand measurable attributes of objects and the units, systems, and processes of measurement | • make decisions about units and scales that are appropriate for problem situations involving measurement. |
| Apply appropriate techniques, tools, and formulas to determine measurements | • analyze precision, accuracy, and approximate error in measurement situations;<br><br>• understand and use formulas for the area, surface area, and volume of geometric figures, including cones, spheres, and cylinders;<br><br>• apply informal concepts of successive approximation, upper and lower bounds, and limit in measurement situations;<br><br>• use unit analysis to check measurement computations. |

# References

Barbeau, E. J. *Polynomials.* New York: Springer Verlag, 1989.

Barbeau, Edward, and Susan Brown. "Welcome to Our Focus Issue on Algebraic Thinking." *Mathematics Teacher* 90 (February 1997): 84–85.

Bass, H. "Algebra with Integrity and Reality." In *The Nature and Role of Algebra in the K–14 Curriculum: Proceedings of a National Symposium, May 27 and 28, 1997,* pp. 9–15. Washington, D.C.: National Academy Press, 1998.

Bednarz, N., C. Kieran, and L. Lee, eds. *Approaches to Algebra: Perspectives for Learning and Teaching.* Dordrecht, The Netherlands: Kluwer Academic Publishers, 1996.

Bransford, John D., Ann L. Brown, and Rodney R. Cocking, eds. *How People Learn: Brain, Mind, Experience, and School.* Washington, D.C.: National Academy Press, 1999.

Breidenbach, Daniel, Ed Dubinsky, Julie Hawks, and Devilyna Nichols. "Development of the Process Conception of Function." *Educational Studies in Mathematics* 23, no. 3 (1992): 247–85.

Burke, Maurice, David Erickson, Johnny W. Lott, and Mindy Obert. *Navigating through Algebra in Grades 9–12. Principles and Standards for School Mathematics* Navigations Series. Reston, Va.: National Council of Teachers of Mathematics, 2001.

Coxford, Arthur F. *The Ideas of Algebra, K–12.* 1988 Yearbook of the National Council of Teachers of Mathematics (NCTM). Reston, Va.: NCTM, 1988.

Cuoco, Al. "Computational Media to Support the Learning and Use of Functions." In *Computers and Exploratory Learning*, edited by Andrea A. diSessa, Celia Hoyles, and Richard Noss, pp. 79–108. New York: Springer Verlag, 1995.

———. "Match Making: Fitting Polynomials to Tables." *Mathematics Teacher* 96 (March 2003): 178–83.

———. *Mathematical Connections: A Companion for Teachers and Others.* Washington, D.C.: Mathematical Association of America, 2005.

Cuoco, Al, and E. Paul Goldenberg. "Regression Lines through Conic Sections." *Mathematics Teacher* 96 (December 2003): 634–38.

Driscoll, M. *Fostering Algebraic Thinking: A Guide for Teachers, Grades 6–10.* Portsmouth, N.H.: Heinemann, 1999.

Dossey, J. "Making Algebra Dynamic and Motivating: A National Challenge." In *The Nature and Role of Algebra in the K–14 Curriculum: Proceedings of a National Symposium, May 27 and 28, 1997,* pp. 17–22. Washington, D.C.: National Academy Press, 1998.

Durell, C. V., and A. Robson. *Advanced Trigonometry.* Dover Publications, 2003.

Education Development Center. *The CME Project.* Boston: Pearson, 2009.

Fey, James T., and Richard A. Good. "Rethinking the Sequence and Priorities of High School Mathematics Curricula." In *The Secondary School Mathematics Curriculum,* 1985 Yearbook of the National Council of Teachers of Mathematics (NCTM), edited by Christian R. Hirsch, pp. 43–52. Reston, Va.: NCTM, 1985.

Goldenberg, E. Paul. "Mathematics, Metaphors, and Human Factors." *Journal of Mathematical Behavior* 7 (September 1988): 135–73.

———. "The Difference between Graphing Software and *Educational* Graphing Software." In *Visualization in Mathematics*, edited by W. Zimmermann and S. Cunningham, pp. 77–86. Washington, D.C.: Mathematical Association of America, 1991.

Good, Phillip I., and James W. Hardin. *Common Errors in Statistics (and How to Avoid Them).* 2nd ed. Hoboken, N.J.: Wiley, 2006.

Graham, Ronald, Donald Knuth, and Oren Patashnik. *Concrete Mathematics.* Boston: Addison Wesley, 1989.

Greenes, Carole E., ed. *Algebra and Algebraic Thinking in School Mathematics.* Seventieth Yearbook of the National Council of Teachers of Mathematics (NCTM). Reston, Va.: NCTM, 2008.

Greenes, Carole, Kyung Yoon Chang, and David Ben-Chaim. "International Survey of High School Students' Understanding of Key Concepts of Linearity." In *Proceedings of the 31st Conference of the International Group for the Psychology of Mathematics Education,* edited by Jeong-Ho Woo, Hee-Chan Lew, Kyo-Sik Park, and Dong-Yeop Seo, pp. 2-273–2-280. Seoul, South Korea, 2007.

Howe, Roger. "From Arithmetic to Algebra." *Mathematics Bulletin* (forthcoming).

Kaput, James J., Maria L. Blanton, and Luis Moreno. "Algebra from a Symbolization Point of View." In *Algebra in the Early Grades,* edited by James J. Kaput, David W. Carraher, and Maria L. Blanton, pp. 19–51. New York: Lawrence Erlbaum Associates, 2008.

Kieran, Carolyn. "The Core of Algebra: Reflections on Its Main Activities." In *The Future of the Teaching and Learning of Algebra: The 12th ICMI Study,* edited by Kaye Stacey, Helen Chuck, and Margaret Kendal. Dordrecht, The Netherlands: Kluwer Academic Publishers, 2004.

Kilpatrick, Jeremy, W. Gary Martin, and Deborah Schifter. *A Research Companion to "Principles and Standards for School Mathematics."* Reston, Va.: National Council of Teachers of Mathematics, 2003.

Lester, Frank K., Jr., ed. *Second Handbook of Research on Mathematics Teaching and Learning.* A Project of the National Council of Teachers of Mathematics (NCTM). Reston, Va.: NCTM; Charlotte, N.C.: Information Age Publishing, 2007.

Mathematical Association of America (MAA). *Algebra: Gateway to a Technological Future.* Report on the "Algebra: Gateway to a Technological Future" Conference, edited by Victor J. Katz. Washington, D.C.: MAA, 2007.

Mathematical Sciences Research Institute. Critical Issues in Education Workshop: Teaching and Learning of Algebra. Workshop organized by Al Cuoco, Deborah Ball, Hyman Bass, Herb Clemens, James Fey, Megan Franke, Roger Howe, Alan Schoenfeld, and Ed Silver, May 14–16, 2008. http://www.msri.org/calendar/workshops/WorkshopInfo/454/show_workshop.

National Council of Teachers of Mathematics (NCTM). *Curriculum and Evaluation Standards for School Mathematics.* Reston, Va.: NCTM, 1989.

———. *Principles and Standards for School Mathematics.* Reston, Va.: NCTM, 2000.

———. *Curriculum Focal Points for Prekindergarten through Grade 8 Mathematics: A Quest for Coherence.* Reston, Va.: NCTM, 2006.

———. *Mathematics Teaching Today.* 2nd ed. Updated, revised version, edited by Tami S. Martin, of *Professional Standards for Teaching Mathematics* (1991). Reston, Va.: NCTM, 2007.

———. *Focus in High School Mathematics: Reasoning and Sense Making.* Reston, Va.: NCTM, 2009.

National Mathematics Advisory Panel. *Foundations for Success: The Final Report of the National Mathematics Advisory Panel.* Washington, D.C.: U.S. Department of Education, 2008.

Piaget, Jean. *Psychology and Epistemology: Towards a Theory of Knowledge.* London: Penguin, 1972.

Sfard, Anna. "On the Dual Nature of Mathematical Conceptions: Reflections on Processes and Objects as Different Sides of the Same Coin." *Educational Studies in Mathematics* 22 (February 1991): 1–36.

Sfard, Anna, and Liora Linchevski. "The Gains and the Pitfalls of Reification—the Case of Algebra." *Educational Studies in Mathematics* 26 (March 1994): 191–228.

Slavit, David. "An Alternate Route to the Reification of Function." *Educational Studies in Mathematics* 33 (September 1997): 259–81.

Suzuki, Jiro. "On Coefficients of Cyclotomic Polynomials." *Proceedings of the Japan Academy, Series A, Mathematical Sciences* 63, no. 7 (1987): 279–80.

Thompson, Patrick W. "On Professional Judgment and the National Mathematics Advisory Panel Report: Curricular Content." *Educational Researcher* 37, no. 9 (2008): 582–87.

Usiskin, Zalman. "Conceptions of School Algebra and Uses of Variables." In *The Ideas of Algebra, K–12,* 1988 Yearbook of the National Council of Teachers of Mathematics (NCTM), edited by Arthur F. Coxford, pp. 8–19. Reston, Va.: NCTM, 1988.

Wilf, Herbert S. *Generatingfunctionology.* 2nd ed. San Diego: Academic Press, 1994.